LIGHT OF THE MIND, LIGHT OF THE WORLD

Light of the Mind, Light of the World

Illuminating Science Through Faith

SPENCER A. KLAVAN

For Mark Bauerlein —
See you in Florida!

— *(signature)*

Since 1947
REGNERY
An Imprint of Skyhorse Publishing, Inc.

Regnery books may be purchased in bulk at special discounts for sales
promotion, corporate gifts, fund-raising, or educational purposes.
Special editions can also be created to specifications. For details,
contact the Special Sales Department, Regnery, 307 West 36th Street,
11th Floor, New York, NY 10018 or info@skyhorsepublishing.com.

Regnery® is an imprint of Skyhorse Publishing, Inc.®,
a Delaware corporation.

Visit our website at www.regnery.com.

Please follow our publisher Tony Lyons on Instagram
@tonyisuncertain.

10 9 8 7 6 5 4 3 2 1

Cataloging-in-Publication data on file with the Library of Congress.

ISBN: 978-1-68451-533-2
ebook ISBN: 978-1-5107-8326-3

Cover design by Jonathan Hay
Cover illustration by Cynthia Angulo
Cover photograph from Getty Images

Printed in the United States of America

For my mother, Ellen

. . . blest the Babe,
Nursed in his Mother's arms, who sinks to sleep
Rocked on his Mother's breast; who with his soul
Drinks in the feelings of his Mother's eye!
For him, in one dear Presence, there exists
A virtue which irradiates and exalts
Objects through widest intercourse of sense.

By Stephen C. Meyer

Recent surveys performed by Gallup, Pew, Harvard, and others show that young people in particular are suffering from a deep sense of meaninglessness. Many attribute their loss of faith to "the science," which they've been told makes belief in God unreasonable. This makes sense, given that over the last few decades many famous scientists and science popularizers have appointed themselves as celebrity spokesmen for atheism. Richard Dawkins, Lawrence Krauss, Bill Nye, the late Stephen Hawking, Neil deGrasse Tyson, and others have produced popular books and television programs arguing that science renders belief in God unnecessary or implausible. It doesn't take a science degree to understand the implications: we are just collections of atoms in a vast collection of atoms, and there is nothing else.

Spencer Klavan describes those for whom science rules out faith in God as "bereft," having been "taught to theorize their own souls out of existence." It's no wonder that we are living through an epidemic of anxiety and despair. But while the common narrative tells us that theists were mired in their religious superstition until a few luminaries like Isaac Newton lit the way to the truth of materialism, Spencer sees it differently: After millennia of "fits and starts," as it were, Judeo-Christian theism, rooted in the doctrine of creation, gave rise to modern science. Only later did the atomistic, materialistic science that replaced theistic natural philosophy *after* Newton lead man "into exile, away from the lovingly created universe of the Medieval God."

Far from suppressing the development of scientific thought, the doctrine of creation was, in fact, a necessary condition for the rise of modern science. Recognition of the contingency of nature, in all its myriad intricacies, upon the mind of a rational Creator was a necessary foundation for rational inquiry and empirical investigation. The Creator had imbued the creation—the whole universe—with order, and because we humans are made in the image of the Creator, our rational minds ought to be able, through careful observation and reasoning, to understand that order. Indeed, men like Johannes Kepler and Robert Boyle helped to break the hold of Aristotelian "armchair philosophy"—natural philosophy without interrogation of nature itself—by emphasizing that the job of a natural philosopher is to ask not what God *must* have done, but what he actually *did*. This belief that nature was ordered and intelligible was the conceptual framework required for humans to make advances—as we did, not in fits and starts but in leaps and bounds—during the scientific revolution. Kepler wrote that God wanted us to recognize natural laws, and that he created us in his image so that we could share in his thoughts. In other words, humanity held an exalted position in creation. Spencer goes even further, arguing that not only is the universe intelligible, and not only are humans uniquely designed to comprehend it, but the human mind cooperates with the mind of God to make our universe what it is through our participation in His creation.

Spencer sets up his argument with a unique and fascinating account of the development of scientific thought leading up to the current resurgence of theistic science. He begins with the ancients—the Chaldeans, the Greeks, early investigations into the natural world through magic and alchemy, the revival of Aristotelianism in the early Middle Ages, the rise of modern science in a theistic milieu, and then the decline of theism with the rise of scientific materialism.

The story Spencer tells interests me deeply. In many ways it parallels the story I tell in my own work. Like Spencer, I see the history of modern science as a rise, a fall, and a rise. The first rise is the development of modern science within the Judeo-Christian tradition of western Europe from the Middle Ages through the seventeenth century. The fall is the decline of theistic science in the face of materialist theories that seemed to explain our origins and our human nature without recourse to a creator. This is exemplified by what Spencer has called the "hyper-Newtonianism" of characters like Pierre-Simon Laplace, Antoine Lavoisier, Voltaire, and Denis Diderot who, unlike Newton himself, took Newton's mechanics as a replacement for God's agency in the universe. Indeed, I find Spencer's take on Newton—that his theology was integral to his science—a refreshing antidote to the secular trope that science and religion are necessarily in conflict. But by the time Newton published his work, the belief that human reason could completely replace religious belief was already becoming entrenched among the thinkers of the so-called Enlightenment. Already, many elite intellectuals saw no need to make recourse to theism to explain anything about the universe—past, present, or future. As Spencer explains, Newton's laws were seized on and divorced from the context *in* which, and indeed the foundation *on* which, Newton himself had presented them. And so it was in the name of Newton but not at all in his spirit that the vision of a purely material, mechanistic "Newtonian" universe came to be.

When Spencer interviewed me for his podcast, he asked me if there was in fact a period of time during which scientific atheism seemed like the most reasonable hypothesis. I think the late nineteenth century was such a time for elite intellectuals. Laplace seemed to have explained the origin of the solar system, Charles Lyell the origins of the geological features of the earth, and Darwin the origin of species, all without recourse to God. Even the problems of our human nature seemed to be explained by the materialist ideas of Marx and Freud. But I believe that despite the narrative of the New Atheists and the media, the materialist zeitgeist has been on the wane and will ultimately prove but a blip in the history of science.

That's because we are in the middle of another rise in the plot. In my book *Return of the God Hypothesis*, I detail three major scientific advances that have made belief in God entirely reasonable for scientists and nonscientists alike. First, the discovery that the universe is not eternal, but had a beginning. Second, the discovery that we are living in a "Goldilocks Universe" in which the fundamental physical laws and parameters are finely tuned for life. And third, the discovery in molecular biology of an entire realm within each single cell of exquisite, intricately constructed molecular machines—information-based nanotechnology. Each of these discoveries speaks to the necessity of *mind* as a precondition for the existence of our universe.

The relationship between mind and matter—the mind of God, the mind of man, and reality on the smallest scale—is Spencer's central theme and the basis for the intriguing questions he asks about current theories of quantum physics and the consequent breakdown of the "hyper-Newtonian," materialistic view of the world. Can matter exist apart from the mind of God? What would the universe be, apart from the mind and experience of man? Do we search in vain for meaning, or does our very experience of the universe, as creatures in the image of the Creator, confer reality on the universe that it could not otherwise possess? Is science bringing us back to God, and placing us with him at the center of all meaning? Scientists have shown convincingly that our universe was created—that it had a beginning. For Spencer, the next question is this: Is it only with our participation that creation is complete?

A Note on Translations

When I cite works in languages other than English, I usually either write my own translations or combine various existing versions. Modern English-language editions of the relevant texts are cited in endnotes and listed in the bibliography for reference. But readers should be advised that my renditions will often differ in minor particulars from the translations cited.

PART 1

Idols of Metal and Wood

Where no gods are, ghosts prevail.
—Novalis, *Christendom or Europe?*

From Dark to Dawn

T his is a book about how God reveals himself through science and human experience. It is a story about how the natural world once seemed alive with spirit and divine fire, and how it might be starting to seem that way again.

For many, the world has come to look dark and dead—like a machine. There are rumors and threats abroad that it will stay that way, that humanity itself will be discarded or surpassed by its own technological creations. It can feel as if religion is on the ebb, as if humanity is a mistake and God is ancient illusion. But this notion has already become outdated, though we haven't yet fully realized it. The argument of this book is that our latest discoveries about the natural world do not make humanity look irrelevant or God seem obsolete. Just the opposite: the world described by science increasingly looks like the world revealed by faith. The lights are coming back on.

It's also quite possible that this book will exasperate both scientists and theologians alike. Since I am neither a scientist nor a theologian, and since I may be justly accused by each of meddling in their affairs, I suppose I had better explain myself.

The World Order

Anyone who tries to discern the imprint of God's hand upon the heart of man—and by telling a story about the history of *science*, no less—is vulnerable to several accusations of extreme presumption. "Who are you," scientists may ask, "to tell us the meaning of our work? If there are puzzles and contradictions in cosmology, *you* will not be the one to solve them. If on our soaring journey to the breathtaking pinnacle of human knowledge we have kicked up a little dust of confusion and raised a few unanswered questions, *you*—a classicist, an antiquarian, a scholar of bygone things—will not be the one to answer them."

And on one level this objection is perfectly justified. When our most distant pilgrim telescopes send back freshly glittering depictions of some new cosmic hinterland, you won't hear about it first from me. I am not the person who will resolve the quandaries presented by hoary galaxies standing resolutely in what should be the youngest regions of space we can access, waiting like stately druids with a youthful mischief in their ancient eyes. I have nothing like the mathematical or the experimental expertise it will take to sort out that kind of alarming paradox.

But that is not my intention. The point of this book is not to contest or amend any particular scientific discovery: it is to say something about the whole nature and structure of the enterprise, the whole practice of seeking knowledge about the natural world. And that practice has always implicitly assumed a rational structure in nature: the word *cosmos* in Greek just means "order." This was the conviction of the earliest known astronomers and natural philosophers. It was the belief of the Chaldean sky watchers of Babylon, by whose measures we still trace the arc of every circle, and the sages of Miletus who hunted out the principles of matter in the growing and dying of earthly things.

To set out in search of laws that govern nature is to insist, often against all appearances to the contrary, that the convoluted muddle of events as we currently experience them is really just one phase of a regular pattern. Despite the odds, we remain confident that the waves we can see lapping against the shore of our tiny island betray currents and tremors that extend unseen through all the dark depths of the sea, whose rhythms, though staggeringly intricate and varied, arise ultimately from the interaction of a few simple principles. That is the belief that gave the geographer Eratosthenes the gall to think he could calculate the circumference of the earth in the third century BC, despite the fact that he had never left the vicinity of the Mediterranean. It is the same belief that motivated the discovery of Antarctica, no less than the discovery of the Higgs boson. After all our millennia of voyaging, somewhere in our hearts we are still certain that the world has an order.

Implicit in that certainty is another necessary truth, more often unspoken than not: the human mind is not an accident. It is stamped with a certain inescapable structure that gives order to our perceptions, that funnels them into language and textures them with meaning, that discerns in the physical world a character of harmony.[1] That world does not simply appear to us as a hectic concatenation of unrelated parts but as an organic whole: freighted with significance, woven through with cause and consequence, shuddering everywhere with tempting whispers of a grander harmony than we can yet discern.

It is natural to look at the stars and wonder why they move across our field of vision. But what's remarkable is that we expect such questions to have *answers*. We may get the answers wrong. But through trial and error, checking one another as we go, we can also get the answers right. We cannot help but feel that the world should be accountable to our logic. Our way of seeing things feels to us like more than an illusion: it feels like a promise. If it didn't, there would be no point in science or in anything else.

Unless we expect the universe to deliver on our innate expectation of order, we might as well just close up shop right now. On the other hand, if we are going to carry on asking questions like "How do quantum effects relate to gravity?" or "Can a superconductor function at room temperature?" then our actions will continue to demonstrate—however we may verbally protest to the contrary—that we believe our minds reveal truths about an ordered world.

However human consciousness came about, it seems to those who possess it to be capable of extracting meaning from its surroundings. And that meaning, in order to be valid, must be latent in the whole universe. It must be threaded through the grain of all things, networked and embedded into a coherent whole that takes shape under our scrutiny. Which means the world, in turn, must be more than just debris. Order, meaning, harmony: these are more than physical things. They are things we discern in and through the material world, but they are not themselves objects. And so, if the human mind can truly know anything at all—and we believe it can—then when it reaches out beyond itself it must encounter more than matter. It must encounter another mind.

That is the argument of this book. For a while in the recent history of our species it came to seem expedient to consider the world independently of any conscious mind—ours, or anyone else's. It seemed helpful to treat certain quantifiable attributes of things—extension in space, motion over time—as absolutes in their own right, separate altogether from humanity and more truly "real" than "merely" human things like color and texture, desires and dreams, evil and virtue. There emerged in the human imagination an image of a world built up out of tiny unchanging objects, swirling thoughtlessly through an infinite emptiness. It was a picture of the world without a mind.

But from the very beginning it was clear to the most perceptive observers that this picture was at best a convenient fiction. If atoms are going to be tallied up, if distances and volumes are going to be measured, who is going to do the recording? At their most basic level, even the properties we ascribe purely to "matter" come, in part, from us. There is no "counting" things or "measuring" them in a purely material world, for the simple reason that there are no numbers at all. Numbers, distances, time itself—these are categories that consciousness *applies* to matter. They are products of the encounter between the mind and the world.

And if that encounter has any validity at all, then when the mind reaches out to know things, it does not simply meet with the world: it meets with another mind. What we are doing when we calculate, as much as when we paint or fall in love, is drawing something forth from the universe that was implanted there for us. The world is not indifferent to us: it is waiting for us to take on its full shape. It is like a home that sits in darkness until we start a fire in the hearth and take the covers off the furniture. We are what makes it whole.

Ping-Pong Balls and Lightning

The central role of humanity in the world is a philosophical necessity, not an accidental contingency: it will not be changed by new data or called into question by new discoveries. Indeed, it may become increasingly necessary in order to *make* new discoveries. The inescapable fact is that the world is made in communion with *us*: that truth has radical and unavoidable implications for the whole modern way of seeing the world. We have only just begun to grasp them.

It has been hard for us to do so because we have not yet gotten over our habit of thinking in terms of mindless objects. In the back of almost everybody's mind there is a script running, an unexamined set of assumptions that shapes the way we see everything. Those assumptions go something like this: The world is made of very tiny ping pong balls moving through empty space. If each ping pong ball were alone in a vacuum it would move in a perfectly straight line. But since they all jostle each other around, the balls move in a variety of trajectories. If we saw them, they would look random, but actually they are following very complicated geometrical rules based on the collisions among them. Different ping pong balls have different properties and powers, based mostly on their possession of something called "energy," which we picture as a sort of crackling light or electric current. Energy is responsible for when the balls do things other than bounce off one another. Sometimes it causes them to latch onto each other or even fuse into one; it arranges them into the larger structures like cats and rivers that we see around us every day. In other words, it holds the world together.

Which of the various entities described by science corresponds to the ping pong balls in our minds? Are we picturing quarks, electrons, atoms, or even whole molecules? For the sake of our casual assumptions, it doesn't really matter. The answer varies depending on what we are thinking about. When we think about systems like soapy water or infected bloodstreams, we picture little lumpy bunches of ping pong balls (chemical compounds), floating smoothly through a viscous sea of other ping pong balls (a fluid substrate). When we think at a smaller level about, say, nuclear fission, the ping pong balls are parts of the atom—protons, neutrons, electrons, or even things like quarks and bosons—which go whipping out away from each other in the wake of an enormous energy field. The actual identity of the smallest possible component ping pong ball doesn't matter all that much. What matters is that whatever we are picturing is made up of tiny parts which move in predictable and knowable patterns.

This picture of the world—the balls-and-lightning picture—is the working model most people use. In its more technical expressions, it amounts to the thesis that atoms and energy make up the sum total of things. Richard Feynman, one of the greatest physicists of the last century, told a group of freshmen in 1961 that the core of most scientific knowledge could be contained in the following statement: "All things are made of atoms—little particles that move around in perpetual motion, attracting each other when they are a little distance apart, but repelling upon being squeezed into one another."[2] Feynman was famous for waxing poetic about the wonder of atoms, and his public-facing work is a decent summation of how pop metaphysics depicts the world: "I stand at the seashore, alone, and start to think. There are the rushing waves . . . mountains of molecules, each stupidly minding its own business . . . trillions apart . . . yet forming white surf in unison." And then, "Growing in size and complexity . . . living things, masses of atoms, DNA, protein . . . dancing a pattern ever more intricate."[3]

But in order to really see the world as ping-pong balls and lightning—or rather, to see that you might already be thinking of it that way—you cannot only listen to scientists. You also have to listen to how ordinary people habitually imitate the things scientists say. The true measure of pop metaphysics is not whether it captures the minds of professionals but whether it conditions the thoughts and everyday speech of the general public. And in this respect the balls-and-lightning theory is unmatched. It has insinuated itself so thoroughly into our thoughts that it now sounds eminently plausible to basically everyone as an exhaustive description of reality. Even those who question it aren't really sure what they might replace it with. "We're all just made of molecules and we're hurtling through space right now," said comedian Sarah Silverman, accepting an Emmy for her HBO special, *We Are Miracles*. In the title routine of the special itself, Silverman announced: "There's no God. I believe in miracles though. Obviously, they're science based. . . . Think about this: every single person in this room tonight . . . there was a time in history, a blip ago in the scope of history, where we were all microscopic specks."[4]

Silverman's Emmy speech is often cited online as a pithy summation of the most profound wisdom our civilization has to offer, emblazoned on inspirational Instagram placards or quoted in appreciative tweets. In fact, the internet is a great archive of amateur philosophizing, and aphorisms like Silverman's have considerable currency online. They are also embedded into popular fiction, from Marvel movies to streaming miniseries. "There is no me," says a dying woman at the end of writer-director Mike Flanagan's Netflix thriller *Midnight Mass*. "There never was. The electrons of my body mingle and dance with the electrons of the ground below me and the air I'm no longer breathing."[5] One concise expression of a similar sentiment comes from filmmaker Rolf de Heer's 1993 cult favorite *Bad Boy Bubby*. Norman Kaye, playing a character called "the Scientist," tells the title character that "we're all just complicated arrangements of atoms and subatomic particles—we don't live. But our atoms do move about in such a way as to give us identity and consciousness. We don't die; our atoms just rearrange themselves. There is no God."[6] You can find parts of this quote sampled on album tracks and discussed minutely on Reddit forums.

It's a conspicuous fact that Silverman and the Scientist were both concerned to disabuse their listeners of apprehension about divine judgment. Endorsements of ball-and-lightning physics are often, though not always, motivated by similar aims. One of the theory's most successful depictions in literature is that of Philip Pullman, the "anti–C. S. Lewis," in the magisterial trilogy *His Dark Materials*. The third installment in Pullman's series, *The Amber Spyglass*, dramatizes a cataclysmic struggle between the forces of enlightenment and those of regressive theocratic control. God, or the being who presumes to pose as him, is a fraud masquerading as the maker of a universe that is really self-composed out of particulate matter: "He was never the creator. . . . He was formed of Dust as we are, and Dust is only a name for what happens when matter begins to understand itself. Matter loves matter. It seeks to know more about itself, and Dust is formed."[7] A universe that explodes automatically into being from the unthinking encounter of blunt objects: this myth—and it is a kind of myth, even if informed by science—expresses reverence for all that can supposedly happen without any superhuman intention. Ideas about what the universe is made of are always, at least implicitly, ideas about who did or didn't make it.

They are also, and relatedly, claims about what *we* are made of. When we talk about the world, we talk about ourselves, because we are in the world. If the universe is made of ping pong balls and lightning, then so are we. Feynman's vision of interweaving atoms made its way into celebrity physicist Carl Sagan's iconic 1980 television series *Cosmos*. "We are made of star-stuff," Sagan was famous for saying. In *The Cosmic Connection* (1973), he wrote, "All of the rocky and metallic material we stand on, the iron in our blood, the calcium in our teeth, the carbon in our genes were produced billions of years ago in the interior of a red giant star."[8] The awe inspired by this notion gives popular appeal to similar claims by Sagan's modern analogue, Neil deGrasse Tyson: "Ever look up at night and feel small?" asked Tyson in a hugely popular Tweet. "Don't. Instead feel large. Atoms in our bodies trace to the remnants of exploded stars. We are Stardust. We are alive in the universe. And the universe is alive within us."[9]

And so, the ping pong-balls-and-lightning idea comes along with a picture of the human body as an elaborate organic machine, built up from the ground out of raw materials. The engine of this machine is the brain, around which the rest is assembled like a protective casing. "The skeleton isn't inside you: you're the brain so you're inside the skeleton"—in 2016 an anonymous user posted this assertion on a massive Reddit forum called r/Showerthoughts (31 million members as of May 2024).[10] The sentiment captivated readers and became a meme: "You're piloting a bone mech using meat armor" is one frequent iteration. "Does the brain control you, or do you control the brain?" asked Karl Pilkington of his fellow comedians on *The Ricky Gervais Show*. "You *are* the brain," Gervais replied, waving away Pilkington's observation that sometimes his consciousness seems more mysterious than the meat-suit theory can explain.[11]

Whatever doubts may occur, whatever complexity we may admit in moments of philosophical abstraction, the touchstone to which we return is the chemistry set in the meat sack. That is the picture we can't get out of our heads, the reflexive way we think about ourselves. "My body, a chemistry lab made of meat, simply chooses to make me feel a little bit nervous for no reason," tweeted "james" (to a response of 434,000 likes as of May 2024).[12] These are witticisms, but the implication is supposed to be that they're funny because they're true: Silverman's law, that we are all atoms hurtling through space, resonates easily with a twenty-first-century audience. In our imagination we move through the teeming architecture of an atomic universe, imperceptibly glittering with cascades of electric dust. Faster than thought, a billion billion granules of existence flow around us and through us to form a world that throbs with conscious life. We never have to think about this notion directly. It is just the backdrop of how we move through the world, a mythology we take for granted as fact. Mindlessly colliding from time beyond reckoning, tiny but impenetrably solid chunks of pure being have latticed themselves into ice and fire, flesh and bone, chemicals and thoughts. That is what the world is, and what we are.

My objective in this book will be to argue that this picture of the world is grievously wrong. The chemistry set in the meat sack moving through a balls-and-lightning world is not simply an oversimplified or crude vulgar caricature of a more dignified scientific truth. Rather, the science itself on which we base our pop imagery is seriously out of date. Atomic physics has been transformed, not to say exploded, by the quantum revolution of the last hundred years. The discoveries of that revolution have been basically shut out from our working model of the universe, for the simple reason that if we let them in, they would bring the whole edifice tumbling down. The foundation is already rotting away, though we do our stubborn best to ignore that fact. Still, it becomes increasingly evident that the tiny solid grains of existence we had pictured are neither solid nor exclusively granular. A house built on sand, indeed.

The point of what comes next will be to trace the history of how our current picture of the world came into focus, to show why it is wrong and in what ways, and then to suggest how we might replace it with something that conforms more readily both to our intuitions and to what science is discovering. It's a matter of urgency that we do so. For the wonderous optimism of our imagined atomic universe

has a way of abandoning its disciples when issues of real consequence come into view. "We're all just atoms and nothing matters
in this universe." "We're all atoms. What's the point of living?"
"What is the point in living if we are made of atoms and everything
around us is an illusion?" "What is the point of living if we are going
to die and not remember anything?"[13] These are the moans of unease
that ripple over message boards when the bright distractions of the
daily news feed have subsided. In the watches of the night, when the
carnivalesque flood stream of online opinions and entertainment
dies down, troubled minds listen for a voice of comfort. And what
they hear is not Carl Sagan's bright professions of enthusiasm, but
fatal silence from a universe that never answers because it never
could. For all its lightning and complexity, at bottom, the merely
atomic universe is less than uncaring, because it cannot think.
When our minds are on more immediately pressing things, we can
ignore this. But in the few crucial moments when the heart cries out
for a depth to meet its own, it becomes a matter of great consequence
to know what the world is made of.

Existential questions, even when we don't recognize them as
such, call for existential answers. So, for instance, when you reach
out in a moment of sorrow for counsel from someone you consider
wise, what will the remedy imply about who you are? Every year
more people seek help with depression or anxiety and receive a pre-
scription for medication to clear the fog that has settled over them.[14]
The question is not whether such medication might be a useful tool
in some of these cases. It's what doctor and patient alike understand
themselves to be doing, what kind of material they think they are
handling when they calibrate the specs of a human soul. The "chemical imbalance theory" of depression, which holds that persistent
ennui is largely a matter of off-kilter brain chemicals, always sat

uneasily with responsible practitioners. Its widespread acceptance was more a matter of convenience for drug salesmen than of genuine scientific conviction. But as a marketing campaign it was ruthlessly effective, in part because of how neatly and plausibly it fit into the picture of the world most people had already drawn for themselves. One study in Australia found that 88.1 percent of respondents "believed a 'chemical imbalance' to be a cause of depression." Find the glitch, take the pill, fix the problem. Chemistry sets, meat sacks.[15]

In *The Haunted Man and the Ghost's Bargain*, Charles Dickens tells the story of a chemistry professor confronted with a supernatural offer to erase his worst memories. Tempted, he asks, "if there were poison in my body, should I not, possessed of antidotes and knowledge how to use them, use them? If there be poison in my mind, and through this fearful shadow I can cast it out, shall I not cast it out?" If we are atoms among atoms, endowed by chance with the strange prerogative of reconfiguring ourselves, then the ghost's diabolical offer becomes ever more attractive. The world is a dead thing, we think, and we are the only known minds in it, alone in the cavernous halls of an empty cathedral with no architect. We have needs and desires, however randomly we came by them. There follows one simple imperative: mold the clay before us into the shape of our choosing. This includes every corner of our environment, but of course it also includes our fellow man and ourselves. Already there are proposals on the table to edit bad memories out of the brain, to make people shorter, to blot out the sun, to reconfigure future generations in conformity with fashionable ambitions.[16]

"As far as we know," writes Yuval Noah Harari in *Homo Deus*, all of human civilization "resulted from a few small changes in the Sapiens DNA and a slight rewiring of the Sapiens brain." Now, somewhat rashly in Harari's view, a new vanguard of scientists dreams of forcing a new evolutionary leap: "maybe a few additional changes to our genome and another rewiring of our brain will suffice to launch a second cognitive revolution." In *The Transhumanist Manifesto,* Dr. Natasha Vita-More writes frankly about seeking "alternative options for perceptual, cognitive, and physical bodies" that conform more readily to our fondest ideals.[17] The new world will be custom-made by the masters of the old one.[18]

People who talk this way are demonstrating the courage of their convictions: things really are just atoms and energy. Crude techniques like hormone injection and surgical implants, honed by repeated use in the mania for "gender transition," will give way to bionic upgrades, Neuralink, and human-machine interfaces that do much more than simply sculpt a man into the image of a woman. The captains of industry propose to make all things new, again and again, at a whim.

The developments of history and the ingenuity of great minds have placed into human hands an awesome and terrible power: the power to transform the world. It is no longer a matter of merely academic interest, if it ever was, to find out *what we are working with.* Before you remodel a house, you examine the foundations; before you perform surgery on a living patient you study human anatomy. By the time we are elbow-deep in the stuff of the universe it will be too late. For the great secret, the cosmic joke, is that whatever we can do to the world we can do to ourselves. Whatever surgery we perform, we will turn around one day to discover it was always us on the operating table. Some caution is advised.

But though this era is perilous, it is also pivotal. The sixteenth- and seventeenth-century events that we now know as the "scientific revolution" didn't materialize out of nowhere. They were steps on a natural path of progression that began as soon as men began to wonder about causes and effects in the natural order of things—a progression toward greater clarity and sharper precision in technical knowledge, but also toward confusion and disillusionment in spiritual matters. The more we learn about the mechanics of life, the less we seem to know about its purpose; the more we can do, the less we seem to understand about why we should do anything. Scientific knowledge has been a matchless source of prosperity, longevity, and power for those who know how to wield it. But trailing in the wake of those achievements has come a deadening philosophy, jokingly referred to by some as The Science™ and more seriously defined by others as Scientism—the bone-deep and unreflecting assumption that matter alone can account for all reality, that all of existence is measurable and automatic.[19] The corrosive effects of this poisonous philosophy can be seen everywhere around us.

Still, the loss of faith and direction that accompanied the rise of science in the West was not the result of a sinister plot or an act of intellectual sabotage. It was a natural stage in the development of humanity, a kind of growing pain we went through. We didn't get here by accident, but we also didn't get here out of sheer malice or willful apostasy. The dark night of the soul came on us gradually; many of the greatest minds along the way resisted it with all their might. But still it came, like an inescapable shadow following in the light of our discoveries, and maybe that was bound to happen. Whether it was or not, I am going to argue that now is the time to shake off that spiritual sleep—for the night is far spent and the day is at hand. If the development of science once tempted us to view the world as a machine, new discoveries are now calling us out of that passing illusion, into a fuller and more beautiful truth. To those with eyes to see, the world described by science is now looking more and more like the God-ordained universe revealed in our ancient scriptures.

To make that case, this book will tell the story of how we got here and where we might go. Part 1 (Idols of Metal and Wood) begins with ancient natural philosophy and ends with the birth of classical physics, tracing the incremental path that led us to suspect the universe might work like a machine. Part 2 (The Fallen Tower) tells how this mechanical idea of the world captured the West's leading minds and its general public alike. But then, at the height of its powers, when its adherents seemed poised to conquer all of space and time, the materialist philosophy inspired by classical mechanics began to fall apart, buckling under the weight of its own contradictions until it was finally shattered by the revelations of quantum physics. Part 3 (All Things That Were Made) traces how those revelations might point the way to a nobler view of mankind and a rich new understanding of the ancient truths of the Bible, reconciling the discoveries of science with the essential wisdom of faith. Dark though the errors of materialism have been, evil though its wages may yet be, we can hope that in history's retrospect they will look like nothing but a passing shadow. Weeping may last the night. But joy comes with the day.

CHAPTER 1

Ghosts in Exile

The general public are under the misapprehension that those who do research into astronomy, and the other arts of necessity that come along with it, become atheists. For astronomers, they suppose, observe things happening by necessity, without any conscious intention or thought directed at an ultimate good or purpose.

—Plato, *Laws* 12.967a

From the groaning depths of an eternal ocean, the world was born. That was the *genesis*, the origin and source of the universe, according to the wise man Thales. His home was the city of Miletus, on the western shore of what is now Turkey. In those days—at the turn of the sixth century BC—the Ionian Greeks of this region sat astride a border between two worlds. To the east was the glittering kingdom of Lydia, storehouse of priceless treasures and object of every conqueror's desire. To the west were the Greek-speaking cities, a complex extended family of competing cultures divided by the dialects of a common tongue. And all of them—mankind with its many races, the patches of earth and rock they fought over, even perhaps the gods they worshipped—sprang forth from water, the womb of all things.

It may have been the poets who suggested this idea. In those days, singers chanted stories for hours at a stretch, layering myth upon myth into an ornate depiction of reality since time began. The grandest of these compositions were handed down by the self-professed disciples of a blind visionary named Homer, who was supposed to have lived a couple of centuries before Thales. In Homer's *Iliad* there are faint suggestions—glimpses only—of a river called Oceanos "whose streams are the source of everything."[1]

Or maybe the story was older still, making its way westward from the plains beyond Lydia. There lay the domains of Assyria and Babylon, competing empires whose kings had stretched out their hands to impose divine order over all the earth. When Thales was born, Babylon was on the rise, shaking off Assyrian dictatorship to reclaim an ancient legacy of world domination. The scribes and stargazers of the neo-Babylonian empire, known as the Chaldeans, toiled away in lavishly funded research institutes to map out the framework of space and time.

The precision of Chaldean mathematics was unrivaled. Their base-60 number system was responsible for dividing the year into twelve months and the circle into 360 degrees, a legacy that endures to this day. But Babylon's researchers handed something else down, too: a solemn story of how the world was made. "When the skies above were nameless and the earth was not yet named, Apsu, the first, who sired them, and Tiamat, the maker who gave them birth, mingled their waters."[2]

The soldiers and salesmen who passed through Lydia brought more than treasure with them. Thales, out on the fringes of the eastern kingdoms, might have heard about the mingling of Apsu and Tiamat's many waters at the font of existence. Snatches of cryptic song from Homer, research briefs from the Babylonian laboratories . . . there were all these teasing hints about an order governing the whole chaotic landscape, some insight marvelous in its simplicity that could reveal what held the world together.

And then there was the evidence of Thales's own two eyes. Out west stretched the great expanse of the Mediterranean, on which the very ground beneath him seemed to float. From the dizzy heights of heaven itself came still more water to drench the earth, and out of that wet soil the plants drew their silent life. It all fit together: the whole teeming biome, with its raucous variety and complexity, arose out of the motions and permutations of this one fundamental substance, this *archē* or "governing principle." The properties of water, and the laws of its behavior, explained everything. Thales could see it happening all around him.

This, at least, is how one interpreter accounted for Thales's convictions.[3] Looking back at the origins of Greek philosophy from its eventual headquarters in Athens, Aristotle of Stagira surveyed the views of his predecessors about what the world is made of. It was an issue of prime importance, a question that occurred as soon as men began to wonder about things. And in the fourth century BC, when Aristotle wrote, there was no more consensus on the subject than there had been in the time of Thales himself. For it was not obvious to everyone that water was existence at its simplest.

Another Milesian, Anaximenes, thought air was more fundamental than water. Heraclitus of Ephesus disagreed: fire, with its searing power of raw energy, must be the catalyst that set the world in motion. And that was just the monists, who thought the world could be reduced to a single underlying material.[4] After them came Empedocles of Acragas, according to whom fire, air, and water were all fundamental—as was a fourth component part, earth. Maybe these four elements, in varying mixtures and permutations, were enough to make everything.

As Aristotle observed, many of these fractious disputants had one thing in common: they "thought that the sources of everything belonged only to the category of matter." To them the world was made of *hylē*, a Greek word meaning just plain "stuff." But this presented a fiendish problem: what makes the stuff move? If some simple substance transforms and mutates into all the other things we see around us, "Why does this happen? What is the cause?" Water, sitting placid in a pool alone, doesn't spontaneously generate activity and life. A mere heap of stuff doesn't suddenly get up and walk. It doesn't even move: it just sits there. Something else must be at work.

Making Things Move

This "something else"—the thing in the world that is more than matter—nagged at the minds of philosophers like a half-remembered dream. What was it? How could anybody know about it? After all, we experience the world through something that the Greeks called *aisthēsis*: sense perception. And by definition, our senses make physical contact with physical things. How, then, could anyone perceive what was beyond the reach of the senses? How could anything more than matter be known?

Aristotle's teacher, Plato, stressed the issue acutely in his dialogues. These were literary dramas about the life of the mind. In many of them, Plato's own teacher, Socrates, gnaws relentlessly at the question of what exists besides "the things you can touch and see and perceive by the other senses."[5] In the *Phaedo* this problem becomes painfully urgent as Socrates awaits his own state-imposed suicide. When the poison slithers its way through his nervous system and jams the signals to his brain, when his eyes go dark and his whole body goes numb . . . what will be left of him?

Plato's answer, presented through Socrates, is that we ourselves are more than flesh. In and through our senses, as our bodies make material contact with the world of physical objects, our minds catch sight of something beyond that world entirely. "Do we proclaim," Socrates asks one of his friends, "that 'the just' is a thing which exists? Or not? . . . What about 'the beautiful' or 'the good?'" It's almost a rhetorical question: we know those things do exist. And yet, "have you ever grasped onto any of them with one of the body's senses?"[6]

You have not. The physical body, the sōma, rubs up against other physical matter. But in perceiving and experiencing that contact, the *psychē*—the mind or the soul—discerns more than mere stuff. It discerns shape and order and beauty. These higher things are only dimly reflected as attributes of the objects we see around us; our eyes see a beautiful sunrise or a beautiful face, not beauty in itself. But the qualities we recognize in matter are eternal, and they are real.[7] In fact they are more real, thought Plato, than the things we can touch and smell and see. With effort we may one day slip from the dead husks of our bodies to see Beauty and Goodness directly, "Not tangled up with flesh and men and things and all the rest of that mortal junk," but absolutely and in themselves.[8]

If Plato thought the world was made of anything, it was made of that: the true, the good, and the beautiful in themselves were what existed for him. Everything else was a dim shadow of what the soul can know. "He gave these things the name 'ideas,'" Aristotle explained: the Greek words *idea* and *eidos*, which we also translate as "Form," suggest something that the soul itself can see. "The things we experience with our senses are derived from these ideas, and named after them."[9]

On this vexed subject, however, there was no more agreement than there was about the generation of the world from water. Plato had taken some cues from certain disciples of Pythagoras, the coy mystic from an island near Thales's homeland called Samos. Pythagoreans "thought that the principles of mathematics were the principles of everything": they reduced the world to the chilly elegance of numbers and equations.[10] The stars wheeling overhead, and the plants Thales saw springing from the ground, were all coordinated in their movements by a few bare numerical postulates. That at least would explain why the human mind could discern not just mere blobs of "stuff," but a knowable order of change and transformation.

Still, were numbers enough to set the world in motion? They could account for planetary orbits. But what about other kinds of change? Did the hearts of lovers really flutter in their chests because the face of the beloved embodied the principle of symmetry? Were soldiers driven from the comfort of their homeland to defend their fathers' gods, simply because they stood for the perfection of the prime numbers? These kinds of motions seemed to spring from an altogether different source.

Instead, perhaps there was a kind of perfection, a totality of being so complete that all things yearned toward it. From a city called Elea on the far coast of Italy, there came to mainland Greece the subtle proposition of a man named Parmenides. For him true reality was total and unchanging, "both the cause of thinking and the thought itself."[11] Parmenides apparently suspected that this meant nothing ever really changed. But Empedocles countered that since things do change, they must do so for a reason—and perhaps that reason was what we call love. It would follow that the attractive power of absolute being, and the hatred of its opposite, motivated every transformation.[12]

Aristotle, though hardly a romantic himself, was very taken with this notion. What if desire drove all things to their determined ends? What if, underlying the bare matter of the universe, there was a longing for perfection that made things move?[13]

Powers of Attraction

Somewhere in this jumble of myth and speculation, amid the babble of debate and the accumulation of human experience, a picture of the truth began to emerge. Sifting through it all, pacing among his students as he lectured them, Aristotle homed in on his theory of the case like a huntsman stalking his prey.[14] Thales and the natural philosophers had seen truly that the stuff of the universe was bound together by some underlying principle. Plato must be right that this principle is something other than the stuff itself, or nothing would ever happen at all. The Pythagoreans insisted that the world beyond matter can be accessed by operations of the mind, as exemplified in the abstract calculations of mathematics. But Parmenides had hit on the crucial sticking point, the paradox whose wrinkles must be smoothed out before it all made sense: how can matter, that confused mess of mere stuff we see and touch, ever come into communion with what is not matter? How can the reality beyond space and time enter into this corruptible world?

Aristotle's answer was revelatory: the two worlds are one. Reality is not divided into forms of immaterial being and lumps of material stuff. Rather, the forms are fused with the matter: everything has a nature that gives it shape and makes it what it is.

The things that are in the world—the *ousiai*, or really existent beings—have both matter (*hylē*) and form or shape (*morphē*).[15] We never just see lumps or blobs of intermingling color and light. For adults with the full use of their faculties, the world does not appear in what psychologist William James would later call "one great blooming buzzing confusion."[16] Instead, from the moment we see or touch anything, our experience is laden with form and distinction. Our vision is already organized by the shapes and patterns that the disciples of Pythagoras traced in their geometry. Plato was right about this much: the raw data of our senses is only the beginning. Sight is more than mere color, and sound is more than mere noise. There is also order. There is also form.

But if matter never sits inert on its own, neither does form ever float free of matter. We can distinguish the form of a bird from the tangle of feathers, bone, and flesh that makes up each particular bird. But still, the essence of birdness never stands naked in front of us as a separate thing, even in the mind's eye. Our thoughts "are not images, but they never take place without images": when we contemplate even simple shapes like circles and triangles, we cannot do

so without picturing them somehow, mentally drawing them in ink
or carving them out of bronze. The world is made of form and matter, both at once.[17]

And if matter comes interwoven with form, then it does not have
to sit motionless and inert. It has potential, or *dynamis* in Greek.
This is where we get our word "dynamic": things can change place
and shape because of the possibilities inherent in what they are. A
stone held above the ground has the potential to fall if released; a
human soul in embryo has the potential to grow and learn.
Motion—not just movement through space, but also every kind of
change throughout time—is caused by the inherent drive that shapes
mere matter into existing things.[18] Not all of these movements are
chosen by a living consciousness: a stone may not "want to" fall to
the ground, except by analogy. But the analogy holds good because
it identifies a fundamental law connecting all living and inanimate
beings. The *telos*, the end goal of a natural body, is like a longing to
be complete.[19]

So those vines and trees that Thales saw springing from the
ground were the realization of possibility, a consummation of the
potential that was once bound up in the tight casing of their seeds.
The thing that made them grow (*phyein*) was *physis*, "nature," the
"source of changing and staying the same which exists in a
thing—not just because it happens to, but in and of itself."[20] Here
was the true origin, the true *archē* of the world's order: nature, a
kind of organic growth, drives all things on to fulfill the potential
inherent in what they are. "All natural bodies and masses are moved
in and of themselves . . . for they have their nature, the source of
movement, within themselves."[21]

Drawn irresistibly in the direction they find inscribed into their
very being, things move toward the goal that makes them what they

are. This is the world as Aristotle described it—an interlocking dance of naturally moving things, each one alight with potential and driven to become whole.

Fire and Spirit

Centuries later, men would find themselves living in Aristotle's world.

The rival tribes of Europe, Greeks and barbarians alike, were brought into uneasy communion by the conquering force of the Roman empire. Then in the 300s AD, Rome's own governing authority fell under the sway of a strange new power, its emperor Constantine seduced by a pugnacious Jewish sect whose worshippers claimed to have seen a man get up from the tomb and walk.

As a political entity, the western Roman Empire would soon crumble back into a rubble of competing territories. But time could not erode the West's new fixation with Jesus the "Christ," God's "anointed one," whose power over men's minds was just beginning to make itself felt. No matter how furiously its opponents might try to snuff it out, Christianity "could slip into your ears under the clandestine cover of silent written words."[22] The allure of the new faith, its defenders argued, was not in its strength but in its truth.

Empires might come and go, but "the Word of the Lord stands forever."[23] Europe's cacophony of political powers and philosophical systems was coming gradually under the sway of a new kind of king. His jurisdiction was not constrained by borders: it was reality itself, the total network of existence which His mind had created out of nothing. One word from this all-powerful consciousness had been enough to make light burst forth out of darkness; His decrees had set the limits and foundations of the cosmos, the entire universe of everything that had ever existed.

In the wake of these dramatic revelations, Greek philosophy became newly significant—and newly dangerous. All mythological speculation must be denied outright: there was one God only, and His new followers would not have His world populated by slatternly nymphs and panting satyrs. In the light of the Gospel it was understood that most of humanity's fairytales were products of pitiable ignorance and damnable error.[24]

But perhaps in the rarified atmosphere of certain privileged schools, God in His mercy had afforded some men a dim vision of true light. There were hints of this possibility among the Stoics, for instance, in their reverent worship of the supreme deity they called "Zeus." "Full of Zeus are all the streets and city squares; the sea is full of him, and the harbors too": so wrote the poet Aratus, who had studied under Stoicism's founding master, Zeno of Citium.[25] St. Paul had quoted Aratus back at the Greek philosophers by way of demonstration that Greece's own luminaries had received inklings of the one true God: "'for we are his offspring,' as some of your own poets have said."[26] So maybe Greek wisdom had resources that Christendom could use.

That certainly proved to be the case in the realm of natural philosophy, where the teachings of Aristotle's Lyceum found generations of unexpected new disciples. By the turn of the second millennium AD, as the institutions of Christian Europe settled into their newfound power and stability, curious minds found themselves newly piqued to understand the design of the physical world. Like Babylon's Chaldean astrophysicists, the Christian researchers of the dawning Middle Ages charted out everything they could know about the heavens and the earth—always in the conviction that whatever they found would bear the mark of minute divine care.

New research consortia duly began to rise against the skylines of cities like Oxford, Bologna, and Paris. Founders and members alike were motivated by an irresistible longing to know God's sublime order, in part by studying the works of his hands. The very name of these new institutes testified to a soaring intellectual ambition: the *universitas magistrorum et scholarium*, the "fellowship of masters and disciples," was meant to be a gathering place for students and teachers from the far-flung corners of the world. The newly minted "universities" of Europe would assemble the best representatives of humanity's many arts and sciences, synchronizing their disparate endeavors and binding human knowledge into a glorious universal scheme. For these talented scholars, wandering in awe through a perfectly ordered world, there could hardly have been a better guide than Aristotle.[27]

"Spirit is the first agent of all. It causes and intends the forms and motions of sublunary bodies. The heavenly spheres are its instrument." So wrote Thomas Aquinas, Aristotle's greatest expositor, who moved among the scholars of thirteenth-century Paris and Rome, outlining the cosmos.[28] Beyond the orbit of the moon—in the "superlunary sphere"—stars of pure fire wheeled in the perfect circular motion appropriate to their natures. Below, in our "sublunary" realm, things moved in more erratic patterns as dictated by their rougher composition out of earth, water, fire, and air. But all things, as Aristotle had taught, moved by nature toward their determined end.

The whole grand dance—from astral rotation down to the mundane falling of a stone—must have some catalyst behind it, or else nothing would move at all. This was the linchpin holding it all together, Aristotle's answer to the problem of how matter could move.[29] This "unmoved mover" must be an immaterial force, the incorporeal being that Aquinas called "Spirit." It must be all action (*actus purus*), raw force and unmixed reality without any shadow of change. Surely, Aquinas argued, it must be God.[30]

The world envisioned by medieval scholars was an expression of pure intention; it was an image of God's inexorable will as uttered throughout a delicate universe of receptive matter. It was raw power working itself out in the realization of uncountable possibilities. "All things whatsoever have order among themselves, and this is a form that makes the universe resemble God," says the incandescent soul of Beatrice in *The Divine Comedy*, Dante's definitive poetic depiction of the intricately ordered medieval cosmos. Beatrice and the hapless pilgrim narrator soar upward into the unimaginable heights of paradise, buoyed by the one power that drives all motion: "This carries fire up toward the moon; this is the driving force in mortal hearts." In the shimmering halls of creation as pictured by Dante and Aquinas, all things, conscious or unconscious, live and move by the power of "the love that moves the sun and other stars."[31]

The Angels Vanish

Magnificent though this intellectual edifice was, it began to show signs of wear under the scrutiny of university scholars. Aristotle's universe hung together beautifully as an exercise in rational theory. But it did not altogether fit with the observed facts. Discovering this, as natural philosophers gradually did from the fourteenth to the seventeenth century, was like revealing cracks in the walls of reality itself.

The world had come to look like a grand palace of God's creation, its floors richly carpeted with grass and set upon a sure foundation of oceanic depths, holding steady under crystalline domes of moving starlight. Now, though, the inspectors who surveyed this

splendid cathedral found tiny fissures in its molding—little inconsistencies, barely enough to catch the notice of an untrained eye.
These minor blemishes seemed inconsequential enough in themselves. But as a challenge to God's perfection of design, they foretold
a catastrophe of the highest magnitude. If left unexamined, they
would splinter the beams of nature and bring the whole building
down. Either the little questions of the philosophers must find sat-
isfactory answers, or heaven itself must be found wanting.

To begin with, there was the matter of projectiles. It was a fact
of all-too-common experience that an arrow, if released from a taut
bowstring, would go whipping through the air until it met with
resistance from the earth or sank into the flesh of some unfortunate
foot soldier. As a heavy inanimate object, it ought to be drawn
naturally straight downward toward the center of the earth. So,
what kept it moving along its trajectory? In Aristotle's world, there
were only a few possible answers—and each brought with it a
swarm of problems.[32]

The arrow might be driven forward by the displacement of air,
known in Greek as *antiperistasis*. What the medievals called *horror
vacui*, the terror of the void, kept the world bursting with substance
at all points: no stretch of space could ever be left completely empty,
for nature abhors a vacuum. If the arrow threatened to leave a void
behind it after departing from its initial position, perhaps air would
rush in to fill that space and push the arrow forward. Or else the air
behind the arrow, driven on by the same blow of the bowstring that
delivered the shot, would outstrip the projectile's downward motion
and carry it forward until the air settled.

At the University of Paris in the 1300s, a teaching master in the
arts faculty named Jean Buridan was among those wrestling with
the issues inherent in projectile motion. Having risen from humble

means to a position of high respect, he took it upon himself to challenge Aristotle's explanations. "Air, no matter how fast it may move,
is easily divisible," he wrote. Arrows obviously slice *through* the air.
They aren't pushed along through space *by* it, or else they would
move faster with big blunt knobs stuck on the end in place of trian-
gular points. More air displaced should mean faster motion. But the
opposite was obviously the case: sharper arrows flew faster, not
slower.

Besides which, arrows aren't the only things that move without the
continuous application of force. Ships continue to glide through water
after their rowers stop heaving. Mills and spinning tops keep turning
in place, even while they stay in the same spot. Air couldn't possibly
be carrying all these objects about, thought Buridan—something else
must be going on.[33]

If it wasn't something in the air, was it something in the arrow
itself? After all, it was supposed to be an inner principle of change,
a *physis*, that moved things along. But arrows and ships are
man-made objects, not natural bodies. And to suggest that projectiles had spirits of their own, that they flew about on invisible wings
of desire to get where they were going, smacked of pagan fantasy
and fairytale. The philosopher who placed a secret motive spirit
inside the flying arrow risked repopulating the whole world with
those cultic deities that had haunted the woods of Greece and Rome.
God's light had chased the idols and demons from their hiding
places, but now spinning tops and javelins were supposed to have
minds of their own? It wouldn't stand.

Buridan helped point the way out of this tangled maze. Neither
the arrow itself nor the air was responsible for the motion, he argued.
Instead, it was the force of the bowstring that left a kind of mark
upon the arrow and drove it home. The action of the archer, at the

moment when he lets his arrow fly, "impresses upon it a kind of impetus or motive force."[34]

This theory of "impetus"—a precursor to what we now call "momentum"—did not stop at arrows and bowstrings. It reached into the heavens themselves. Up there in the superlunary sphere, stars and planets were supposed to move by the design of supernatural intelligences, agents of God who carried out his will more swiftly than thought. The intuition that the heavens were managed by super-human regents, acting on God's orders, went back at least to Plato.[35] But if the first mover had imparted to the stars an initial force, leaving within them an impetus strong enough to keep them moving, then there was no need for any higher consciousness to push things along. And so, wrote Buridan, "one could imagine that there would be no need to propose intelligences which move the heavenly bodies, since Scripture does not say anywhere that they should be proposed."[36]

Looking back on this astounding insight, it is common to suggest that Jean Buridan chased the angels out of the firmament and the life out of the world. Science historian Richard C. Dales referred to this moment as "the de-animation of the heavens."[37] Herbert Butterfield, in a series of lectures at Cambridge, said that the theory of inertia which eventually developed out of questions like Buridan's "helped to drive the spirits out of the world and opened the way to a universe that ran like a piece of clockwork."[38]

It's true enough that the world was starting to look more mechanical. Buridan's fellow natural philosopher, Nicholas Oresme, would speculate about a universe constructed by a God like a great clockmaker, "making a clock and letting it run and continue its motion by itself."[39] It might seem as if nature was being drained of spirits and purged of angels, its motions reduced to the predictable workings of machinery.

But were the spirits really gone, or just displaced? After all, what exactly was impetus? What, for that matter, was the "force" that sent the arrow flying in the first place? We might ask the same question now of words like "momentum" and "energy." We imagine that by reducing these concepts to clinical terminology, applying words to them derived from the erudite Latin of the Middle Ages, we have rendered them totally material and so lifeless. Perhaps, though, we have just given them colder and less descriptive names.

The ancient pagans imagined gods and heroes racing across the sky in the form of constellations and planets. For our part, we think we have domesticated the gods, stripping them of primitive superstition and revealing them for the mere objects they are. This, we imagine, enables us to bring their motions into regular arrangement and predict the clockwork patterns within them.

But Jupiter and Orion moved just as regularly across the sky when they were associated with characters from myth. And the motive force that inheres in a projectile does not become less mysterious simply because you call it "impetus" instead of "spirit." Jean Buridan identified where the motion of an arrow comes from: he saw that it comes from the application of force. But he said nothing, and claimed nothing, about what the nature of that motion is. Force, impetus, momentum, energy, motion: these are words we use to tame the spirits, because they do not seem to carry associations of life and will. But those associations are there, even if we don't notice them: the words themselves are just dead metaphors, images of some invisible activity.

A "force"—a *vis* in Latin—is an application of strength. Momentum is *movimentum*, the power and fact of movement. All these words carry lost imagery of life and action, and none of them explains where the life itself comes from, or who is acting. The controversial Oxford scholar William of Ockham, confronted with the notion of impetus, objected that it was just another word for the fact of motion itself—a description of a tendency, not a designation of any real entity.[40] In the name of his famous simplicity, Ockham proposed doing away with impetus altogether.

But there it stubbornly remained, just as "force" and "momentum" stay seated in our minds today. What are they? They are surely not objects, and by invoking them we have not reduced the world to mere matter. We may draw them in our mental diagrams as little red arrows or rushing lines of wind, but those are only visual symbols of something that cannot be directly seen, material representations of more than material reality. What we are really doing when we use words like "force" or "impetus" is painting a mental picture of something more than material. We are describing things that we can neither see nor touch directly, though we experience them acting in the world nonetheless.

We imagine that we have chased the spirits out of the world. But in reality, they have become ghosts in exile, haunting the edges of our sight. There has always been something at work in the world to make it move, something more than matter which gives life to otherwise inanimate stuff. The poet John Milton wrote of God that "Thousands at his bidding speed / And post o'er land and ocean without rest."[41] Now, instead of unseen winged servants, we picture lifeless energy fields and currents of force. But these exiled ghosts are no more purely material than the angels were, and the world has not been purged of life just because we call the spirits by new names. The project of reducing our living universe to a rational order would generate marvelous insight. But it would also cut us off from our essential connection with things, setting humanity on a path into exile and alienation. We are only now, after hundreds of years, coming to the end of that path. But it was at the close of the medieval era that we embarked on it. The crisis was just beginning.

CHAPTER 2

Fixing the Stars

This to attain, whether Heaven move or Earth,
Imports not, if thou reckon right; the rest
From Man or Angel the great Architect
Did wisely to conceal, and not divulge
His secrets to be scanned by them who ought
Rather admire; or, if they list to try
Conjecture, he his fabric of the Heavens
Hath left to their disputes, perhaps to move
His laughter at their quaint opinions wide
Hereafter; when they come to model Heaven
And calculate the stars, how they will wield
The mighty frame; how build, unbuild, contrive
To save appearances; how gird the sphere
With centric and eccentric scribbled o'er,
Cycle and epicycle, orb in orb . . .

—John Milton, *Paradise Lost* 8.70–84

A s the spirits hid themselves from view, medieval philosophers began to suspect that the world ran like clockwork. Men like Nicholas Oresme had hinted that the universe might be mechanical, its motions set running by a divine inventor and then left to whir methodically along in perfectly regular order.[1] If so, then a breathtaking possibility presented itself, as intoxicatingly attractive as it was perilously close to blasphemy: perhaps the workings of the heavens could be known and predicted like those of any other machine. And maybe, just maybe, mankind could seize the levers of control. It was not the first time the thought had occurred. The dream of a mechanical cosmos was an ancient dream.

In 1901, off the rocky coast of a Greek island called Antikythera, a team of unsuspecting sponge divers laid hands on the oldest known mechanical model of the heavens.[2] The "antikythera mechanism" dates from around the second or first century BC, the turbulent final centuries of the Roman republic. For two thousand years it had moldered away at the bottom of the sea, hidden in the ruins of a sunken ship. All that's left of it are a few notched dials and some bronze gears, but the purpose of the original design is clear enough. As the gears turned, mechanical pointers would crank into place to indicate the position of the sun and perhaps even the planets, computing the position of each heavenly body among the stars on a given night.[3] This was the first known attempt to build a metal universe.

The Roman statesman Cicero, a voracious polymath who prided himself on rubbing shoulders with the best and brightest in every discipline, made a point of mentioning that the Stoic researcher Posidonius had constructed just such a miniature model of the heavens. It was a magnificent triumph of information technology, "a sphere which, with each singular rotation, transfers the same motion to the sun and the moon and the five other planets." Today we call this kind of planetary model an "orrery," after the earl whose patronage helped support the creation of a similar machine based on the modern picture of the universe. But that wasn't until the eighteenth century AD: Posidonius was wildly ahead of his time. According to the Stoics, his ingenious device furnished self-evident proof that the universe was constructed by a divine mind. Any human being, even those remote primitives in a far-off land called Britain, would naturally recognize "that the model was constructed by rational design."[4]

In fact, though, Posidonius's marvelous gadget was not self-explanatory. It might have been met abroad with bewilderment as well as wonder. The notches on its face would have indicated the zodiac, a Babylonian division of the night sky into twelve regions associated with twelve major constellations. One pointer would have shown where the sun was positioned each day relative to those constellations, marking its gradual yearly journey along a line called the "ecliptic." Each planet also charted its own path across the background of the "fixed" stars, so called because they all seemed to move together across the sky at unchanging distances from each other. Against the backdrop of this stellar orb, a few lone bodies appeared to "wander" (*planān* in Greek), wending along the solitary paths that earned them the name *planētai*: "wanderers," or "planets."[5]

The resulting map of the heavens is well known to astronomers across the world today. But it would have been rather less than obvious to those far-flung Britons of Cicero's imagination. They would never have heard of the zodiac, let alone the ecliptic. If some enterprising traveler really had brought Posidonius's orrery along on an expedition beyond Rome's borders, he would have had to explain in some detail just what the treasured object meant. If he said it was a model of the universe, his bemused hosts might have gently inquired whether all Romans thought the sky was made of bronze.

In response, the ambassador would have to explain that the device was not meant to convey that particular aspect of the universe. Pointing from the metal cogs to the glittering occupants of the night sky, he would have to translate the symbols of the orrery into the observed positioning of the stars and planets above. The moving parts were not intended to be made of the same stuff as the stars, but to indicate their positions in relation to one another. In other words, Posidonius's device was not simply a reproduction of the universe in miniature. It was a representation of external realities, a visual language for conveying and predicting certain observed experiences.

It could hardly have been otherwise. After all, there was nothing to go on but the appearance of objects in the night sky as seen from the ground. These "appearances"—in Greek, the *phainomena*—had been the source of all astronomical data since the days of Babylon. And ever since those days, the project of the astronomer had been to discover patterns in the changing sky. If these patterns were regular, then their development could be anticipated: what had happened before would happen again. This belief motivated the priests who looked for omens in eclipses to predict the outcomes of battles, no less than the mathematicians who built scale models of the night sky to foretell the positioning of Venus. Finding the order in things meant knowing the future.[6]

Saving the Appearances

The challenge, then, was to find the simplest and most reliable description of how the stars moved across the human field of vision. Plato threw down the gauntlet for generations to come: "He posed the following problem for mathematicians," wrote the erudite analyst Simplicius in the sixth century AD. "What orderly, uniform, and circular motions may be proposed to preserve the appearances of the planets?" [7]

The key point was in that last phrase: the circular motions of the planets were *proposed* (*hypotithemenoi*) to *preserve the appearances* (*diasōthēnai ta phainomena*) of the wandering bodies. A "hypothesis" in this original Greek sense is a supposition, a foundational premise that gets "laid down beneath" the structure of a mathematical model. It is not an observation but a framework that draws observations into harmony with one another—a mental scaffolding to build upon.

The hypothesis of circular movements—perfectly regular, mathematically satisfying, pristine in their simplicity—was the tool used to predict where the planets would appear in the cleanest and most precise possible way. The circles that astronomers drew across their astral maps were like the arrows we draw on ocean maps to indicate how the tides move: if you go to swim in the Atlantic, you will not find literal arrows floating in the water. But you may feel the water moving in the direction indicated by the arrows. The visual language of the map, rightly interpreted, can predict your experience of the thing itself.

In this same way, the wheeling models of the ancient astronomers were not hard and solid replicas of the real spheres. They were visual representations of something beyond the material world, depictions of a higher and purer reality whose nature could only be known indirectly by its effects on the senses. The planets themselves, those glittering images that could be seen moving across the night sky, were nothing more than sensory glimmers of an invisible truth. They were aftereffects of a heavenly motion that left its imprint glowing on the retina, like spots of light on camera film.

"These pinpoints of light in the sky are dotted upon our field of vision" says Plato's Socrates in *The Republic*. "As such, they may certainly be considered the clearest and finest things of that kind. But they fall far short of the truth."[8] The real nature of the world beyond the moon was just dimly sketched by the human mind. Groping its way from appearances as they swam before the eyes, man's reason could arrive at approximations of the underlying logic that determined where the stars would go next.

And so for Plato's most devoted pupils, astronomy just meant writing up hypotheses—mental models that could tell the future. It need hardly trouble them if more than one model did the trick. Plato's student Eudoxus wrote his equations as if each planet was carried around the earth by a series of interlocking spheres, their axes and rotations delicately counterpoised to produce the motions observed.[9] This was the model Aristotle found most appealing, since it kept the universe neatly organized around a single point. But in the second century AD, the Roman mathematician Ptolemy supposed that the planets were not bound to one center: they could be pictured as carried along in loops by the interplay of two or more circles around different points.[10] For most of the observed phenomena, these two sets of equations *both worked equally well*: they accounted for all the known information, and they predicted what would be seen next.[11] For all intents and purposes, then, neither was any more "real" than the other. Both were mere sketches, interchangeable ways of giving shape to the intangible reality beyond the visible realm.

But there was one way in which not all models were created equal. The heavenly bodies did not only move: they also glowed, more brightly at some times than at others. As they loped across the sky, these wanderers could be seen to blaze and dim with variable brilliance. This made it look as if they were moving closer to the earth and farther away. Eudoxus's model couldn't account for that. If all the planets moved around the same point, then they always stayed at a fixed distance from the earth. "Saving appearances" meant lining up a hypothetical model with actual observations, so that mathematical conjecture would intersect neatly with sensory experience. And here was one appearance that was better saved by Ptolemy's description: if the planets were circling around various moving centers, they would close in and shine more brightly in the night sky at various times of the year.[12]

There were still multiple ways of explaining things, as Ptolemy himself well knew. The paths of the planets could be resolved into circles that moved around a point, which itself circled the earth. But that point in turn could be placed in various locations relative to the planet's trajectory.[13] You could move the counters around to different starting positions on the board and keep playing the game. The equations would work out the same either way, and any one of them that churned out the right predictions would be fit to purpose.

Still, Eudoxus's model was definitively inferior, and that raised a tantalizing possibility: perhaps there was some mathematical picture that would work better than any other, once and for all. Maybe some map of the heavens would predict everything, and then it could be counted on as *real*—tangible and physical as a set of metal gears. If so, then humanity was not condemned to grope by halting half measures toward a vague and numinous truth. What if truth was solid, and could be touched? What if you really could construct a universe machine?

Aristotle himself might have appreciated this suggestion, impatient as he was with abstract speculation about unobservable realms. But it would take Jean Buridan and Nicholas Oresme to really open up the possibility of an actual cosmic machinery. With the closing centuries of the medieval era came the hint that objects on earth moved through the air according to set trajectories—and so perhaps the planets were governed by a knowable law as well. The hunt was on for *true* models, for concrete and physically accurate descriptions of how things moved in heaven and on earth. The men who took up the chase would become famous ever afterward for bringing the universe to heel. Their names were Copernicus and Galileo.

Wheels Within Wheels

It started quietly enough. In 1543, when he published *De Revolutionibus Orbium Coelestium* (*On the Revolutions of the Heavenly Spheres*), Nicolaus Copernicus was nearing the end of his life. It had not been an especially dramatic one: born in the Polish town of Toruń, on the gently curving banks of the River Vistula, Copernicus obtained a middling administrative post in the church diocese of Warmia before traveling to Italy to earn his doctorate in canon law. John Calvin and Martin Luther both made their names during his lifetime, and the Protestant Reformation would soon set a fearsome blaze of religious dissension tearing through Europe. But Copernicus, for his part, lived and died a meekly faithful servant of the Catholic Church. His dissatisfactions were not theological: they were mathematical.

He was a man of the high Renaissance, through and through. A full suite of arts and sciences was at his fingertips, and they all seemed to mesh beautifully together. As the brilliance of Greek and Roman philosophy lit up the world anew, every fresh insight found its rightful place as an adornment of God's masterfully constructed universe. And this was why Ptolemaic astronomy offended him. He found it ghastly to layer circle after circle upon the heavens, cluttering the sky with geometric crosshatchings that never totally fit the observed patterns of movement they were supposed to describe. At this point the foot soldiers of human knowledge had been gathering data for thousands of years, painstakingly marking down anomalies in the position of each star and planet as it appeared night after night. The records had grown voluminous and incongruous enough that no simple model could possibly accommodate them. Every scribbler with a compass and an astrolabe seemed ready to layer on new circular motions ad hoc. The noble discipline of saving the appearances had devolved into a desperate and never-ending patch-up job.

It was "as if someone took hands and feet, and a head and other pieces, from various places—each of them perfectly well depicted, perhaps, but not for the purpose of representing a single person. Since these fragments would not belong to one another in the slightest, what was put together from them would be a monster, not a man." So Copernicus wrote to Pope Paul III in the preface of *De Revolutionibus*, explaining why he found the theories on offer so maddeningly inadequate. "The philosophers could by no means agree on any one certain theory of the mechanism of the universe, crafted by a supremely good and orderly creator."[14]

To a certain extent, this was shop talk: the incongruities of stellar and solar motion would only offend an eye whose vision had been honed to detect computational infelicities. "Mathematics is for mathematicians," wrote Copernicus. On another level, though, cleaning up the astronomers' mess was a matter of grave public concern. The disjointed monstrosity of inelegant physics was an insult to God, whose work should exhibit the perfection of pure simplicity. It was in this respect that Copernicus advertised the marvelous advantages of turning the problem on its head: everything could be made cleaner and clearer by a single change in perspective. The trick was not to explain the sun's motion, because it wasn't moving at all. The earth was.

The Central Flame

From there, a whole new firmament fell awesomely into place. At the center of all things was not earth but fire: not the solid rock of this sin-heavy world but a blazing furnace of divine light and heat, around which all the other bodies danced. "Geocentrism," according to which earth was at the center of things, had been the majority view for all of recorded history. But the "heliocentric" universe, with the fire of the sun throbbing at its core, was not Copernicus's invention. Aristarchus of Samos had hit upon the same idea in the third century BC. So had certain followers of Pythagoras who, as Copernicus pointed out, pictured the earth spinning around a luminous "central flame."[15]

Like the Pythagoreans, Copernicus thought this vision had more than calculational efficiency to recommend it: heliocentrism was also beautiful, and theologically dignified. Among the educated set of Renaissance Italy, where Copernicus had traveled to earn his degree, one could find certain enlightened sages waxing lyrical about the superiority of fire over earth as a material emblem of God. "Nothing reveals the nature of the Good more entirely than light," mused the Neoplatonist Marsilio Ficino. "The sun can signify God himself to you."[16] In Book I of De Revolutionibus, Copernicus burst into his own rapturous peroration on the same theme: "In the middle of it all, the sun sits enthroned. Could we place this luminary in any better position, from which he can illuminate the whole at once? He is rightly called the Lamp, the Mind, the Ruler of the Universe."[17]

But Copernicus also took an important tactical lesson from the Pythagoreans: know your audience. Pythagoras was famous for presenting the general public with only those arguments that would keep them from poking further into matters they couldn't understand. Mobs, he knew, were never more dangerous than when threatened by unsettling knowledge. So, he reserved the really sacred mysteries for the few listeners who had ears to hear. Among those chosen students—the *esōterikoi*, or those "within" the inner circle—esoteric truths could be revealed.

Copernicus had a major advantage in playing this double game: his equations, framed by abstruse and convoluted Latin glosses, were too elaborate for all but initiates to comprehend. He could lay his esoteric teachings right out in the open without fear of broadcasting them too widely. Even in plain view, they would really be visible only to specialists. Mathematics is for mathematicians.[18]

The later sections of *De Revolutionibus* presented heliocentrism not as a dazzling cosmic spectacle, but as an intricate system of calculation. Those who combed through the figures and equations would find the planets' apparent motions explained as products of humanity's ever-shifting viewpoint: earth itself was a planet, and its motion helped explain why the other heavenly bodies seemed to travel in such erratic patterns.

For instance, a moving earth could simplify the explanation of "retrograde" motion, the times of the year when planets seemed to reverse their normal course and travel briefly in the opposite of their accustomed direction across the sky. In principle, at least, this could be accounted for if both the earth and the other planets were traveling around the sun: sometimes earth would "outstrip" another planet for a time, passing it by as both of them moved in the same direction.[19] These kinds of ideas could potentially reduce Ptolemy's innumerable cycles to a cleaner circuit, one befitting the genius of a divine mind.

In practice, however, Copernicus's system ended up just as convoluted as Ptolemy's, with just as many added circles to account for weirdness in the data. There was inaccuracy in the recorded observations of his predecessors, and it would be another generation before Johannes Kepler realized that the circle itself was not the shape that best fit the planets' apparent trajectories. For the time being, experts went on puzzling over how to make sense of the heavenly motions. It was not self-evident that Copernicus had the best answer. At his death, it seemed as if all he had offered was a model among models, one of various possible ways to save the appearances. Heliocentrism or no, the sun would go on rising in the east. Or it would seem to do so—and, since it was all a matter of appearances, what was the difference?

Sunspotting

Beneath the calm exterior of these academic discussions, a revolution was brewing. In truth, Copernicus was not just playing around with numbers or speculating about the symbolic properties of light. He had proposed that the universe had a real, tangible arrangement, and that the sun was quite literally at its center. Andreas Osiander, a Lutheran priest who oversaw the publication of De Revolutionibus, tried to smooth things over by appending a preface in which he laid out more modest ambitions for the book: a central sun was just a hypothesis, one of many possible mental models for predicting what would be seen. "These hypotheses need not be true or even probable. To the contrary, if they provide a calculus consistent with the observations, that alone is enough."

For a while this might have conciliated any skeptics who would otherwise have raised an eyebrow at Copernicus's audacious proposal to set the earth spinning. But Osiander was protesting too much. The promise of *De Revolutionibus* was not merely to save the appearances but to set forth the truth of things in themselves, as they are. Knowledgeable readers would realize this, and they would not be so delicate about it as Copernicus had been. What he had murmured in complex mathematical Latin, others would shout from the rooftops in plain terms that everyone could understand. None would do so more loudly than Galileo Galilei.

In 1609, while he was still working in relative obscurity as a professor at the University of Padua, Galileo heard news that an optician in the Dutch provinces had successfully arranged two glass lenses in such a way as to magnify objects at a distance. That was all he had to go on—just a report of the thing's existence. No prototype, no specs, no prior expertise in lens grinding. But this meager information was enough to set him laboring away in his chambers, until he emerged triumphant with his own *cannocchiale*: a "lens-tube," or—to use the more refined name it acquired from Greek—a telescope.

Not everyone instantly understood the urgency of such a contraption. But Galileo soon made it clear what he was up to. In 1610 he published *The Starry Messenger* (*Sidereus Nuncius*), a short pamphlet on his preliminary findings. Flush from his visual journey into the highest heavens, the intrepid astronomer hurtled breathlessly back to earth with revelations from on high. His ambition was nothing less than "to have put an end to the debate about the galaxy. . . and to have made its essence manifest to the senses as well as the intellect."[20] By furnishing images of previously unknowable objects, the far-seeing eye would penetrate the very heart of nature and reveal her as she really was. That is what Galileo thought the telescope could do.

Without a doubt the new invention could facilitate more observations than ever before, revealing to the senses things previously
accessible only by theoretical guesswork. Galileo saw craters on the
surface of the moon and dark blemishes on the sun itself, suggesting
that the heavenly bodies were not so smoothly rounded as philosophers had imagined. He saw planets that behaved like moons, and
moons that orbited planets. Through his wondrous new lenses,
Galileo watched as satellite bodies shuffled around Jupiter, and
Venus passed into phases of partial shadow.

This last observation in particular seemed to him like evidence
that Copernicus was right: Venus must be orbiting around the sun,
bathed in light on only one side at any given time. The bright side
of the planet would then face toward earth on some nights and away
on others—but only if both bodies were in orbit around the source
of light. Galileo was convinced. Despite some initial Pythagorean
shyness, he was encouraged by the job security of a plum new post
at the court of Tuscany's Grand Duke Cosimo II de' Medici. So, by
1613, in a collection of *Letters on Sunspots*, he could venture a preliminary conclusion: accumulating evidence seemed to make it plain
that the earth really did revolve around the sun.

Matters of Perspective

Quite unlike Copernicus's bookish Latin treatise, *Letters on Sunspots*
was a bold salvo in the vulgar tongue of Italian, maximally accessible
and ripe for scandalous public discussion. A battle royale ensued,
as both Galileo and the clerics who attacked him staked out increasingly uncompromising positions. To begin with, there were scriptural objections to be raised: passages in the Hebrew Bible described
the sun "making its circuit from one end of the heavens to the other"
(Psalm 19:4–6) and stopping miraculously still in its trajectory

(Joshua 10: 12–15) over an earth which "can never be moved" (Psalm 104:5). But these problems were not insurmountable. Back in the sixteenth century, a Spanish hermit named Diego de Zúñiga had observed in his *Commentary on Job* that even the most devoted Copernicans "frequently refer to the course of the earth as the course of the sun."[21]

In other words, the Bible was written from an earthly perspective, describing how things looked to men on this moving planet. Divinely inspired though the scripture was, God had condescended to reveal things as seen through human eyes. Thomas Aquinas had already made much the same point, and Nicholas Oresme had noted that *if* the earth were moving, it would nevertheless appear motionless to mankind, "as if a person were on a ship moving east very swiftly, without being aware of the movement."[22] Copernicus himself had quoted the Roman poet Virgil, whose hero Aeneas watched from the stern of his own ship as "the dry land and cities backed away."[23] Obviously, Virgil was well aware that harbors don't move; boats do. But like the authors of scripture, he had described things as they appeared to his protagonist. It was all a manner of speaking.

When drawn into theological debate, Galileo was perfectly capable of making these and other points besides: following cues from no less an authority than Saint Augustine, he argued that "the holy Bible can never utter untruth. . . . But it is often very abstruse, and may say things which are quite different from what its bare words signify."[24] This might have been enough to quiet his critics when it came to the scripture. But when it came to the physics, Galileo insisted that the earth's rotation was *not* a manner of speaking or a matter of perspective. It was the actual fact of the case.

This was the real sticking point. Galileo well and truly thought that he had "made the essence of the universe manifest to the senses."

He said so again and again. The whole universe was made of real, tangible bodies, and geometry could enable a true description of those bodies as they moved through space. The telescope had pierced beyond Plato's curtain of darkness, beyond mere appearances and symbolic language, to a mathematics of the reality of things.

Alarmed by these claims, the church hierarchy assigned Cardinal Robert Bellarmine the unenviable task of investigating them. Bellarmine—author of a famous Catholic catechism, a future "doctor of the Church," canonized as a saint in 1930—was not an unreasonable man. He took pride in keeping up to date on new theories, though whether he grasped them as exactly as he fancied was a matter of awkward dispute. In any case, he was suspicious of mathematical speculation, and in 1616 he instructed Galileo with the full authority of the church "that the doctrine attributed to Copernicus, that the Earth moves around the Sun and that the Sun is stationary in the center of the world and does not move from east to west, is contrary to the Holy Scriptures and therefore cannot be defended or held."[25] Neither Bellarmine nor Galileo's friend Maffeo Barberini, the future Pope Urban VIII, would hold Galileo back from discussing heliocentrism as one among a number of hypothetical alternatives—as long as it turned out to be the wrong one. *De Revolutionibus* was placed on a list of prohibited books, as was the *Commentary* of Diego De Zúñiga. It was now forbidden to affirm that Copernicus was right on the facts.

This was the one thing Galileo could not abide. He was sure that numbers and measurements offered the most definite truth possible, a kind of instruction manual for the machinery of the world: "Philosophy is written in this grand book, the universe, which stands continually open to our gaze," he declared in 1623. "It is written in the language of mathematics, and its characters are triangles, circles, and other geometric figures."[26] There it was in plain terms: the world was made of material objects, and the bare facts about those objects were to be found in their mathematical properties.

In this conviction Galileo drew a profound division in the mind of man. It would become known as the distinction between "primary" and "secondary" properties. The characteristics of the things we see around us are of two kinds, Galileo claimed. One kind can be described in mathematical terms: bodies moving through space have shape, size, position, quantity, and motion. These facts are *really true* about those bodies, which would have their primary characteristics whether anyone was there to observe them or not. If a book is on a table and not under it, if it is six inches wide and not seven, and if there is one copy of it rather than five or two, these are objective facts, thought Galileo. They are independent of the human mind.

But all the other qualities we experience in things—secondary properties like color, taste, and smell—are only experiences. They are products of the human mind, born in the communion between us and the solid things of the world. When it comes to the things in themselves, wrote Galileo, these subjective qualities are "nothing but empty names." If we vanished from the earth, then so would they.

Clockwork

Outside of us there is a world made of real and solid objects. They are as mathematics describes them. That is the bedrock of everything as Galileo saw it, the primary reality that exists no matter what.

When the Inquisition put Galileo under house arrest in 1633, this view of things was one of the main counts against him. "The author claims to discuss a mathematical hypothesis, but he gives it physical reality," the final report intoned.[27] Pope Urban had tried to dissuade him from this line of thought, suggesting that man's mind could only ever grasp probable truths about creation—it was always possible that the real workings of the universe were concealed in some inscrutable design of God's.

But in his final offending work, the *Dialogue Concerning the Two Chief World Systems*, Galileo made a mockery of that idea. His spokesman in the dialogue, Salvati, drives the point home: "although the celestial appearances might be saved by means of assumptions essentially false in nature, it would be very much better to derive them from true suppositions."[28] Galileo went to his grave broken but not repentant—in his heart of hearts, he never stopped believing that he had seen the truth of things.

Had he? Galileo's own arguments for his point of view were frequently shaky or outright false—he thought, for example, that the ebb and flow of the tides were driven by momentum imparted to the sea as the earth moved. And in the words of historian Thomas Kuhn, "the telescope argued much, but it proved nothing."[29] Evidence that seemed incontrovertible to Galileo often turned out to have an entirely different explanation. The case was not so cut-and-dried as he insisted it was.

And yet, just like the spheres of Eudoxus, Ptolemy's geocentric universe was fraying around the edges. In time, more observations would emerge that couldn't possibly have been made from atop a stationary earth. By 1838, stellar parallax—the hairline slippage in our field of vision as the planet carries us marginally closer or farther away from a given star—could be seen through a telescope. The heliocentric model predicted this; the geocentric one didn't. In that sense, at least, Galileo was right.

But his deeper claim—that mathematics was a literal description of concrete objects as they moved mechanically through space—was far from proven. The success of heliocentrism as a tool for describing and predicting human experience made it look as if this might at last be the real model, the metal universe of Posidonius given final form. After Galileo, a gleaming new picture of the world came into focus.

But it would turn out, at last, to be *only* a picture—one that would be refined and ultimately replaced, as Copernicus's own vision of things had replaced Ptolemy's. One day particles of light and electricity would seem to move in ways no object should, and the supposedly solid world of the Renaissance imagination would start to dissolve. And then an old truth, guessed at by Plato, would make itself felt again: there is *no such thing* as a real machine that can model the world exactly, because the world is not a machine. It is not made of dead objects driven mindlessly through space, and human experience is not some "secondary" illusion, a cloud of fiction layered onto the hard facts of mathematics. In truth our human relationship with the world—our indispensable presence in it—helps give it even its mathematical form. The conscious experience of things, as seen through a living mind, is not simply an irrelevancy to be explained away. From the beginning, when the world was made, it was *seen*.

Today, physics itself is revealing that Galileo's primary qualities—the brute facts of things existing in space—are not after all totally independent of the mind. It is not just things like color and sound that come into being through the relationship between mind and matter. Even looking through a telescope, even flying on metal wings into the heavens themselves, mankind will always, by definition, be seeing appearances. And so perhaps appearances are not so "secondary" after all—perhaps what talk about when we describe the universe is always, can only be, our experience. Perhaps it was meant to be that way.

The things we meet in the world are not dead objects; they are the impression of a moving nature on our living senses. Plato's spots of light, those astral pinpoints on the black canvas of the heavens, might grow brighter and larger as we approach. They might appear in ever-greater detail. But they will always appear *to us*, and it will always be a human eye that sees them.

As Galileo sat in captivity, however, the necessity of these truths was not yet apparent. It seemed to him and his closest associates that the church had cruelly denied him at the crucial moment, when he felt sure he was on the verge of grasping a world that could be handled like a metal object. But someone else would surely come after him to finish the job. For though they had succeeded in imprisoning him, his opponents couldn't keep his picture of the world from taking hold. The ghosts of the cosmos would be banished into exile; the movement of things on heaven and earth would be explained. It would all run automatically, without any conscious intervention.

This was a fantasy, destined never to be realized. But in the years to come its power would only grow. The machine might be a fiction, but it would go on working in the Western mind. Its gears had already been set inexorably in motion.

CHAPTER 3

The Birth of the Small Gods

... it seem'd
A void was made in Nature; all her bonds
Crack'd; and I saw the flaring atom-streams
And torrents of her myriad universe,
Ruining along the illimitable inane,
Fly on to clash together again, and make
Another and another frame of things
For ever:
—Alfred, Lord Tennyson, "Lucretius" ll.36–4

W hat is the world made of? As the seventeenth century unfolded, the stakes of that question rose with alarming speed. Galileo boasted that he could strip Nature herself naked and reveal the true, physical structure of the universe. The Church's reaction showed how seriously its leaders took this claim.

A perilous new intellectual territory was opening up, and those who ventured into it would have to tread very carefully indeed. But what explorer with an ounce of spirit could resist blazing a trail into this enticing terrain? If the dangers were extreme, still the incentive was nothing less than admission into the inner sanctuary of reality itself.

Light in the Darkness

Man's vision of the universe was expanding outward; stars and comets began scattering like spangles beyond the bounded spheres of the medieval imagination. But that did not mean things had to dissolve into chaos—just the opposite, in fact. The hope was that the world would become bigger *and* more regular at once, an ever-vaster field of space reduced to ever-simpler rules of motion. The most devoted astronomers were convinced that God's creation was answerable to human reason. No benevolent creator would willingly mislead his creatures. With the right application of a few clean and clear rules, everything would hang together. That was an article of faith.

No one delivered on the prophecies of this faith more spectacularly than Johannes Kepler. While Galileo's controversy was still brewing, Kepler was gathering his own observations in the German and Austrian territories of the Holy Roman Empire. All over that troubled realm, the storm clouds of religious war were gathering.

Amid the turmoil, Kepler dodged from city to city in pursuit of clarity from beyond and above the human world.

All around him, an unbearable tension was mounting. Citizens of the Empire were weighing the audacious pronouncements of Calvin and Luther. Battle lines had begun to spiderweb across the continent. Princes and fiefdoms lined up on opposing sides of a fateful new question: would the Catholic Church continue to serve as God's arbiter of absolute truth? Or would each monarch, and perhaps even each believer, chart a separate path?

In 1618, a pitiless war broke out that would not end until the newly splintered nations of Europe had given up all hope of uniting under the banner of one church. But even though the political unity of Christendom was destined to collapse in the bloody fratricide of what came to be known as the Thirty Years' War, Kepler's mathematical universe was falling into a glorious formation. As he pored over reams of archival data and new astronomical observations, insights struck him like sunbeams bursting down from heaven upon the gloomy Earth.

The planets, argued Kepler, moved not in circles but in ellipses. These elongated forms were concentrated around two focal points, one of which was the sun. That was the first law of their motion. The second was this: as a planet swept along its path, an imaginary line between it and the sun would carve out equal chunks of the ellipse in equal amounts of time—with the result that the planet would speed up as it moved closer to its fiery anchor, and slow down as it moved farther away. When these two rules were observed, a smooth and satisfying order came to rule over celestial phenomena that had once seemed hopelessly jumbled.[1]

If there was no unity on earth, at least there was composure in the heavens: faith in divine simplicity would be rewarded in the end.

The year after the war began, Kepler published his third law, comparing the orbits of multiple planets to one another and finding that each exhibited the same relationship between its rate of travel and average distance from the sun. The book in which he demonstrated these marvelous consonances was called *Harmonices Mundi: The Harmony of the World*. The dominion of men might crumble into anarchy. But the angelic choirs of the stars would sing in tune.[2]

The foundation of their harmony, and therefore the center of the universe, was unquestionably the sun. This was a fact, not a convenient mathematical fiction: "In astronomy as in every other science, the conclusions we teach the reader are offered to him in all seriousness," wrote Kepler. "Mere plays on words are excluded."[3] For him the sun, and the sun only, made sense at the center. The mathematics it enabled was too magnificently simple to be false.

One could go further, thought Kepler: the sun brought all things into such perfect balance that it was "worthy to become the home of God himself, not to say the first mover."[4] Its perfection radiated outward into the farthest reaches of space and time, governing a staggeringly enormous universe down to the last cubic millimeter. From its white-hot heart in the sun's core, reason itself held the fabric of the world together.

Enchantment

So it was that the study of nature grew more exhilarating even as it grew riskier. A word out of turn or an impolitic overstatement could bring dire consequences. But for those who dared to venture the effort, physics took on the character of sacred revelation. Here was a flaming sword to slice through the veil of ignorance that separated the race of men from the dwelling place of God. When it worked, it worked like magic.

In fact, it *was* magic, in its noblest and most potent form. That was the view of Baron Verulam, Viscount St. Alban, better known to history as Sir Francis Bacon. Bacon acknowledged that much of what passed for magic was just the trickery of deluded cranks. But that was not what he meant by "magick": there was another, more "honourable meaning" that he hoped to secure for the word. Real magic, he wrote, was "the science which applies the knowledge of hidden forms to the production of wonderful operations."[5]

This was not mere pseudoscience: this was science itself. The promise of magic and the promise of natural science had always been one and the same.[6] Both proposed to identify which practices, which combinations of materials and ideas, would reliably produce predictable results. Merlin's spells and the incantations of Dr. Faustus were supposed to tap into secret tendencies in the logic of things: say these words, mix these ingredients, get this result. The premise of natural science was nearly identical: apply this serum, solve this equation, and the following consequences will emerge.

And so alongside astronomy there was also astrology, which Kepler himself practiced until the day he died. Alongside mathematics there was also divination, and each purported to predict what would happen tomorrow based on signs observed today. Know the logic of cause and effect and you know the future; know the future, and you can control it. That was always the ideal, of natural science and magic both alike. The difference was that the science worked.[7]

The task was not to abandon magic but to siphon off and refine those magical practices that could produce real results, deliver real power. In this aspiration, under the favorable auspices produced in England's court by the ascension of the Protestant King James, Bacon laid out what he considered the ground rules for doing magic right. The result was the *Novum Organum*, a new handbook of reason and knowledge to replace Aristotle's outdated operating system.

Bacon's intent was to "open and establish a new and certain course for the mind," beginning "from the first actual perceptions of the senses themselves." Casting away the "idols of the mind"—the predispositions of nature and nurture that prevented most men from seeing things accurately—good researchers would commune instead with the living God. His truths would be revealed directly to mankind's senses, through the bare evidence of "direct experience" or, in Greek, *empeiria*. Once the sound practitioner had seen things "empirically," without the interference of any predetermined theory, then he could try explaining what he saw.[8]

Bacon's followers would look for "true hypotheses": not simply working models for predicting how things would look, but real explanations that would reveal the tidy logic hidden beneath scrambled masses of recorded observation.[9] If multiple competing hypotheses suggested themselves, experiments could be designed to judge between them. Once things were boiled down this way, the tendencies of nature could be revealed and harnessed at their simplest. The potential of this method, if it worked, was limitless. Magic was only the beginning. Their new tools in hand, natural philosophers could even deliver on the solemn promises of an ancient and mysterious craft: they could at last unlock the secrets of alchemy.

The God's-Eye View

"We will show the end of this our art: / An end most useful and most quickly learned, / For nothing strange it needs save that one stock / From which all things by nature are produced."[10] These verses, composed by an unknown sorcerer in the latter half of the first millennium AD, laid forth the alchemist's promise. Everything in nature came from "one stock," a shapeless "prime matter" that could be endlessly transmuted and recombined to produce every substance and phenomenon in the world. Like some primordial clay in the hands of God, prime matter could be molded into anything at all.

Alchemists hoped that this marvelous substance could be obtained with sufficiently deft application of such basic conditions as pressure and heat.[11] Like pure ore pouring forth from baser metals in a smelting furnace, prime matter would come oozing from the belly of existence at the command of whichever genius found the right technique. Ancient legend had it that initiates could learn to accomplish this "great work," the *Magnum Opus*, from the "thrice-great" sacred scribe of Greece and Egypt, Hermes Trismegistus. The coded hints of an apocryphal "Emerald Tablet," where the principles of alchemy were supposed to be inscribed, teased readers about "the miracles of one thing." Whoever grasped this fundamental substance could manipulate the raw materials of creation.[12]

For those who pursued it, prime matter was an object of ecstatic veneration. It was the marrow of being, scrubbed clean of all its accidental characteristics, simple as it had been at the dawn of time before it suffered the touch of fallible human hands. It could be used not only to draw gold from lead, but even to create the legendary "philosopher's stone." This fabled talisman had the potential to turn dead tissue into living flesh and roll back the curse of sin that had been laid on the world since Eden.[13] Prime matter could give its possessor eternal life and power to transform the world.

It had never been found. The hunt of the alchemists, the long hours spent hunched over crucibles and beakers, blanching the shells of hen's eggs until they turned snow-white . . . it was all in vain. Or at least it had been, until men like Bacon and Galileo ushered in the age of the mechanical universe with its mathematically objective "primary qualities." These pioneers felt they really could boil down the world, paring away irrelevancies and unveiling the numerical logic of things as they truly were, in their simplest and most powerful state. "Bacon, like Moses, led us forth at last," wrote the poet Abraham Cowley. After ages of wandering fitfully in the deserts of confusion, mankind had reached "the blest promis'd land" of true enlightenment.[14] Now, finally, the rudiments of existence could be discovered.[15]

In 1660, a new "Royal Society for improving natural knowledge" began gathering in London to compile and examine masses of empirical data—a testament to Bacon's legacy. Unlikely researchers brought in findings from every colony of Britain's dominion and every level of her motley society. Sailors came tramping in from the shipyards with nautical observations. Merchants furnished samples of exotic flora and fauna from Asia and the Americas. The ferocious turmoil of religious civil war having exhausted itself, Europe unleashed its energy instead on exploration and discovery. Natural science became a less recondite and suspicious affair: its advocates had demonstrated its bona fides, and now everyone from kings to common boatswains was interested to see what these purveyors of new knowledge could do.

But what they still could *not* do, at least not yet, was see beneath the surface of the things they studied. The world machine was ticking along nicely, and its workings were apparently as regular as one could wish. But those in the business of manipulating that machine wanted to peer under the hood and get their hands on the gears; they wanted to see not just how matter behaved but what matter *was*, in the pure and essential form that the alchemists sought. And that was a trickier task.

For it turned out, to the dismay of Bacon's heirs, that empirical observation was never *only* empirical.[16] It always came saturated with ideas about what was being observed. Neither you nor I can have a "pure" sense experience unmediated by any interpretation. I don't simply see patches of black and white and brown in front of me as I type this: I see shapes and objects, in this case a table with a laptop on it. If I think for a moment that the table is made of wood rather than synthetic plastic, then I am mistaken in my interpretation of what I experience, much as I might misread a word on my screen. I can replace my false interpretation with a new one. But I cannot simply have *no* interpretation, however hard I try.

Theory is not just derived straight from a direct encounter with matter, or from what the members of the Royal Society learned to call "matters of fact." Ideas always color observations like a pair of glasses that you can't take off. You can change the lenses, but there is no removing the things altogether. Bacon himself had ruminated on this very fact. "On waxen tablets you cannot write anything new until you rub out the old," he wrote. But "with the mind it is not so; there you cannot rub out the old till you have written in the new."[17]

If things are only known through the senses, then those things must have some property that *affects* the senses—some color or shape or texture, some impact on a measuring device, that makes them perceptible. But this means the mind will always "get in the way" of matter. We can never access it in its supposedly truest form, without any "subjective" qualities that depend on a human subject doing the perceiving. So how could anyone, ever, obtain any truly "objective" facts about things as objects—the third-person, God's-eye view of matter in itself?

Men had longed for centuries to answer that question. Now they seemed to have come so close. But though they piled up fact after fact in the hungry archives of their growing empires, still the forbidden knowledge eluded them: there was apparently no route through the senses to the heart of things.

Facts and Figures

Perhaps another approach would do the trick. If Bacon had tried in vain to climb from raw sense data to true ideas, the French luminary René Descartes would move in the other direction, descending from the high abstraction of pure ideas to the facts of things as they really are.

In 1633, when Galileo was imprisoned by the Inquisition, Descartes was thirty-seven. He was on the verge of publishing a minute analysis of how creation hangs together, suggestively titled *The World, or: A Treatise on Light*. But in his alarm at Galileo's imprisonment, Descartes held back. *The World*, he later wrote, dealt with "a number of questions currently in dispute among the learned, in which I have no desire to embroil myself."[18]

As everyone "among the learned" would realize, the questions Descartes meant were about the composition and revolution of the Earth. *The World* would not see publication until years after Descartes's death. But throughout his life he would press on with his efforts to describe reality in its simplest possible form. He was not quite prepared to involve himself in the vexed question of heliocentrism. But he certainly joined Galileo in the effort to identify matter's primary qualities, the objective features that could be expressed in ironclad mathematics.

This effort was not entirely without precedent or pedigree. Already in the fifth century BC, the "laughing philosopher" Democritus of Abdera had announced with impish good cheer that most human experience was simply hogwash. "The surface of things is a matter of convention," Democritus insisted: "it is by convention that things are sweet or bitter." In fact, the whole varied panoply of color and sound, taste and texture, amounted to nothing more than a shifting shadow play. The true origin of these sensory experiences was in the definite motions of solid bodies, so tiny as to be indivisible—or, in Greek, *atomos*. For philosophers like Democritus these "atoms"—minute grains of irreducible being—moved invisibly at the bedrock of reality. They, and the endless void through which they moved, were the only real things in the world.[19]

The ancient atomists had never quite driven their point home, in part because they had not gone far enough: their atoms were still loaded with attributes of their own. Bitter food, they argued, was composed of sharp atoms which pricked the tongue like miniature blades. By contrast, the atoms of sweet food coursed smoothly over the palate. But that just pushed the problem one step further: if bitter atoms were sharp, what gave them the property of "sharpness?" If "sweet" was a matter of mere convention, what did that imply about the "smoothness" of sweet atoms? If atoms explained the qualities of everything, what explained the qualities of atoms?[20]

For Descartes, the answer was to strip away *every* merely subjective decoration from the framework of things, attempting a clean break between mind and matter. The result, he thought, would be a material world knowable in terms of pure facts and figures. Primary qualities, after all, were not really qualities so much as *quantities*: even if things like color and shape were constructed in the human mind, there were other aspects of existence that could be measured with the certainty of numbers. "The nature of matter or body in its universal aspect," wrote Descartes, "does not consist in its being hard, or heavy, or colored, or anything that affects our senses in any other way, but solely in the fact that it is a substance extended in length, breadth, and depth."[21]

Mathematics could be grasped by the mind with unshakable accuracy: one was one and two was two, and neither was even a vanishingly small amount more or less than itself. For Descartes, only "clear and distinct" perception of ideas could guarantee the existence of the thing contemplated—himself included. "I think, therefore I am": this world-famous assertion underwrote an entire Cartesian universe, from the God whose existence could be known by way of his essential nature as pure being, to the facts of his creation when laid forth in the deliciously irrefutable terms of numbers and equations.[22] Extension through space and motion over time: when it came to matter, these two things alone were absolutely and unquestionably real.[23]

The world, thought Descartes, was a *plenum*: a full and unbroken expanse extending outward in all directions like the coordinate system that still bears the name "Cartesian." At every point in the universe was some solid object, and each such object jostled against those around it to move them via a process that is now called "contact action." In *The World*, that unpublishable treatise on things too' controversial to name, Descartes filled his private universe with tangible bodies of pure matter. Like the Greek sages who broke down the infinite variety of things into bare elements of heat or moisture, Descartes proposed that everything was made of fire, air, and earth.

Fire was infinitely divisible, argued Descartes. As such it could furnish the medium through which all things communicated their motion to one another, a kind of "ether" in which even the lightest breath could trigger a chain reaction to move the air around it. Air, in turn, was composed of tiny round pieces "like grains of sand or dust." Tangible and solid chunks of the third element, earth, helped make up the solid bodies of everyday experience. But everything, from the planets in their slingshot orbits to the finest particle of dew, moved in mathematically regular ways through a bursting fullness of space that connected all things, however distantly, to each other.[24]

Bodies and Voids

This kind of thinking, in various forms, was gaining traction. Gradually a new picture of the world was taking shape, and though its details were hotly debated, its outlines were coming inexorably into focus. It was called "the mechanical philosophy": a conviction that the world was made of bodies whose movements and collisions were as sharply regulated as those displayed by an intricately calibrated machine.

And like any machine, the world of moving bodies could be best understood by taking it apart and examining its most basic components. Suddenly the theory of atoms had new appeal. Men like Pierre Gassendi in France and Giordano Bruno in Italy would press the point excitedly, hoping to find in microscopic atoms the "seeds" of existence that gave rise to everything else.[25] If they could be described using numbers alone, atoms might reveal the truth underlying all things.

Even Descartes, though he thought matter was indivisible in principle, still talked in practice of "corpuscles"—a Latin coinage describing the "little bodies" of denser material whose motions could explain the way the world behaved.[26] It was impossible to resist the allure of their purity, these tiny embodiments of unadulterated truth, moving in the manner prescribed for them from on high by the serene logic of mathematics. Whoever could discern the order of their motion, peering behind the haze of merely human vision, could claim to have laid bare the laws that governed the universe.

Posterity would award that laurel to the English physicist Isaac Newton—though he himself would have accepted the credit only with painfully mixed feelings. True enough, his ambition was to achieve a grand synthesis of all movement and change, delivering at last on the hope that one set of laws could describe both planetary orbits and the descent of ripe fruit from trees in autumn. What Newton saw was that all bodies, from the smallest to the largest, pulled toward one another without any mediating help from the ether. Instead they were all of them tugged toward each other by a strange new power that would come to be known as gravity.

As above, so below; on earth as in heaven. Newton's three laws of motion shattered the imaginary barrier once thought to separate the region beneath the moon from the domain of the stars. There was no "superlunar" or "sublunar" realm anymore. It was all one, governed everywhere by the same perfectly consistent laws, just as Hermes Trismegistus was supposed to have intimated. Now physics could illuminate the cryptic utterances of the old alchemical masters. Reason would stretch from high above the clouds down to the mundane realms of daily life with perfect consistency. Man's mind could range across a universe composed at all points out of the same basic materials, according to the same fundamental laws, "to accomplish the miracles of one thing."

Everywhere and always, Newton's first law dictated that "quantities of matter" would retain whatever state of motion they were already in, until some new force altered or arrested their trajectory. When it did, the second law would take effect: the force would cause acceleration, overcoming the inertia of the body affected and changing its rate or direction of motion. The extent of the change was determined by how much force had been applied, divided by the quantity of the matter, or its "mass."[27]

In a sense, forces were absorbed into moving bodies, imbuing them with what Newton would call a *vis insita*—a motive power that soaked into the object itself and gave it the "momentum" that kept it moving. Finally, said the third law, an equal and opposite change would shudder through the body which applied the force, altering its momentum in the inverse direction as far as its mass allowed. Momentum would always be "conserved," meaning its transfer would exhibit the pristine balance of perfect equilibrium. Not a jot of it would be created or destroyed.

After laboring painstakingly over their presentation among his colleagues at the Royal Society, Newton delivered these three laws to the public in the summer of 1687, in a book entitled *Philosophiae Naturalis Principia Mathematica: The Mathematical Principles of Natural Philosophy*. Against a backdrop of "absolute space" and "absolute time," solid objects would move relative to one another in the manner dictated by Newton's equations. The angular momentum of planets came from an attractive force called gravity, exerted on every mass by every other mass. Apples and bowling balls would plummet to the ground when released in midair, drawn by the Earth's bulk until they crashed to the surface and were prevented from moving further. Here, it seemed, was an exhaustive rule book for how things moved.

The Cradle of Nature

But *what* were the things, and what were the "forces" that made them move? These were ancient questions, and they remained unanswered. Perhaps they could be put to rest in the case of objects like billiard balls rolling along a table: large, solid bodies collided with one another simply enough through contact action. But what were the billiard balls made of, and what was this force that pulled them toward the bosom of the earth when dropped from midair? Uneasily, Newton referred to forces of "attraction" inherent in mass itself: "particles . . . are moved by certain active principles, such as is that of gravity," he wrote. Moreover, "God, who gave animals self-motion beyond our understanding, is without doubt able to implant other principles of motion in bodies, which we may understand as little."[28]

In the *Principia*, Newton resolutely avoided speculating about what component parts of matter were acting upon each other and how. "I here design only to give a mathematical notion of those forces, without considering their physical causes and seats," he declared.[29] The whole point of talking about "quantities of matter" was to measure the amount of "stuff" present without getting too specific about what the "stuff" actually was.

Rightly understood, the *Principia* described patterns of motion that could reliably be observed throughout the universe. It did not state what forces and matter were in their essence, "underneath" their appearance to the human mind. When drawn into debate on the topic by his rival Gottfried Leibniz, Newton famously avowed that "I frame no hypotheses. For whatever is not deduc'd from the phaenomena, is to be called an hypothesis; and hypotheses, whether metaphysical or physical, whether of occult qualities or mechanical, have no place in experimental philosophy."[30] How was it that objects could pull at one another without the contact action of the ether? This might simply be the sort of question that physics was not designed to ask or answer. For Newton, after all, it was a matter of urgency to stress that material explanation was altogether insufficient to answer ultimate questions about the fundamental nature of existence.[31]

Whatever could not be observed through experiment must remain a hypothesis, which to Newton meant a speculation about hidden realities underlying the unfolding of events. Newton did not doubt that those realities existed. But he knew he had no evidence to prove what they were. Some of them, up to and including the origins of the universe itself, must forever remain subjects of theological inference: they were by nature inaccessible to material observation and explanation. Based on his research so far, he could say how objects would move. But he would not say what did the moving. Concepts like "force" and "atoms" were convenient ways of describing the patterns revealed by his observations—but none of his experiments proved they were really there.

Yet it was becoming nearly impossible not to believe in them. An early fascination with alchemy had convinced Newton, as he wrote in the first edition of the *Principia*, that "any body can be transformed into another, of whatever kind."[32] Somewhere there were raw materials that had been used to make everything. And it looked ever more likely that they took the form of indivisible building blocks which attracted one another: "we have the whole course of a large experience for the universal gravity of matter and for the hardness of its particles without any instance to the contrary."[33]

Newton himself was adamant that the order he observed in nature must come pouring forth from one mind, which created and sustained the world: "the appearances of things," he wrote, could reveal to human eyes the visible imprint of God's invisible hand.[34] But others would receive Newton's mathematical rules in an entirely different spirit, looking not beyond the laws of nature to their author, but beneath them to a world supposedly composed of solid bodies that ricocheted through an endless void. If these solid bodies had the mysterious ability to attract each other, that only added to their marvelous power. Once their existence was allowed, these invisible beings seemed to explain everything. Their enduring presence underlay all that shifted and changed in human experience, governing the motion of things from the beginning to the end of the world.

The old spirits of movement, those ghosts in exile, the angels that once moved the planets and the stars: all these beings had been hidden behind names like "force" and "momentum." Their life and energy were concealed in the hard casing of abstract words whose implications of passion and intention had been drained away. But what was *vis insita* if not that inner power of motion that the ancients had wondered over? What was gravity if not that driving spirit that Kepler had once called the *anima motrix*, the moving soul that reached out from the sun to call the planets homeward?[35] For centuries the power to create motion out of stillness had belonged to God alone in his capacity as the unmoved mover. But now another motive force was making itself felt, and it called across the void between every particle of mass.

Though he fought mightily against it, Newton's insights would be used in the coming centuries to craft an entirely new depiction of the world. Atoms were taking on a life of their own in the mind of man: they were becoming small gods in their own right. Newton himself had meditated on the possibility that the hidden deities of long-discarded legend were at last revealing themselves, transfigured in the light of science into their true forms. "All things, that is the four elements, are born from this one thing, which is our Chaos," he wrote in a paraphrase of alchemical theory. "The earth is the nurse, Latona washed and cleansed, whom the Egyptians surely had for the nurse of Diana and Apollo," Newton went on: unadulterated, rudimentary matter was "the source of the perfection of the whole world."[36]

To Newton it was beyond question that there was one ultimate craftsman who ordained this perfection, a single God from whom all subordinate authority must descend. But another interpretation was lurking, one that would wrest power from on high and place it solely in the hands of the alluring new deities of the natural world. "They say then that Love was the most ancient of all the gods," wrote Francis Bacon, hinting at this new mythology. Now philosophers could pull the cloak of superstition off of Cupid himself to show what cosmic attraction really meant: "The fable relates to the cradle and infancy of nature, and pierces deep. This Love I understand to be the appetite or instinct of primal matter; or to speak more plainly, *the natural motion of the atom;* which is indeed the original and unique force that constitutes and fashions all things out of matter."[37]

Newton pointed the way up from the laws of nature toward the mind from which they must have sprung. But an altogether different path would open up in subsequent generations. The path that led man into exile, away from the lovingly created universe of the medieval God, would bring him now into a cosmos dominated by new powers, mindless and uncaring but wondrously reliable. The old unruly deities of ancient times would be replaced by atoms, whose operations could be understood and even controlled. This was the birth of the small gods, the tiny bodies and invisible forces that would come to move through the human imagination in eternal and unbreakably regular patterns. Peering intently at their clockwork motion, scientists could hope that one day they might even rule over these mighty entities, the sons of man elevated to the status of lords among deities. The small gods were unseen but awesomely powerful, capable of exerting their will throughout the unending regions of space. They were everlasting, they were perfectly rational, and above all, they behaved. The energy that moved among them was enough to govern the world.

PART 2

The Fallen Tower

The aim and ideal of all natural science is a materialism wholly carried into effect. We here recognize this as obviously impossible, confirming another truth that will result from our further consideration: all science in the real sense ... never aims at the inmost nature of the world. It can never get beyond the representation.

—Arthur Schopenhauer, *The World as Will and Representation*, volume 1, section 7

CHAPTER 4

The Frozen World

"Why do you doubt your senses?"
"Because," said Scrooge, "a little thing affects them. A slight
disorder of the stomach makes them cheats. You may be an
undigested bit of beef, a blot of mustard, a crumb of cheese,
a fragment of an underdone potato. There's more of gravy
than of grave about you, whatever you are!"
                    —Charles Dickens, *A Christmas Carol*, Stave One:
                                                    Marley's Ghost

W hen it came to the small gods, no man was more devout than Pierre-Simon, Marquis de Laplace, the great astrophysicist of eighteenth-century France. Laplace's faith was simple: the laws of physics could account for all things. "A consciousness which, in a given instant, knew all the forces by which nature is animated, and the respective positions of the entities that compose it, could encompass in one formula the movements of the largest bodies in the universe and those of the smallest atom." So wrote Laplace in his famous *Philosophical Essay on Probabilities*.[1] The specter of this inerrant mind had first appeared to haunt Laplace in 1773, when he was in his twenties. He would spend his whole career outlining the contours of the world as they might appear to the all-seeing eyes of such a flawless intellect.[2]

It became known as "Laplace's demon," this perfect consciousness under whose gaze the intricate totality of things would lie frozen in an instant. For such a mind, "nothing would be uncertain. The future, like the past, would be present to its eyes." Human beings, pitifully limited in their capacities, might be struck dumb by the sheer number of bodies and the baffling complexity of unfolding events. But a superhuman mind, with enough processing power to grasp the right initial conditions, could know all time and space down to the slightest tremor of the smallest particle. This inexhaustible kind of knowledge seemed possible to Laplace, at least in theory, with the power of Newtonian mechanics.

Of course, the demon was in the details. In an idealized system, with two bodies isolated in the safety of the imagination, the orbital motion of celestial bodies could be described regularly enough. But the whole staggering premise of classical mechanics was that *nothing* existed in isolation *from anything else.*

Not only did the sun exert its gravitational pull on the planets: the planets also tugged at one another. This mutual attraction had an array of eerie and alarming results, just one of which was that Jupiter's orbit apparently contracted gradually inward while Saturn's expanded ever outward. If this state of affairs were allowed to persist, the machinery of life would eventually come shuddering to a calamitous halt.

Laplace would have none of that. In 1773 at the Académie des sciences (France's equivalent of the Royal Society), he demonstrated mathematically how variations in planetary motion would even out over time on their own.[3] For a man so young—for anyone, really—this was a staggering achievement. It seemed to vindicate the idea of an automatic universe, one that could run without the daily ministrations of a divine repairman.[4] Laplace would claim for Newton's laws a dominion far vaster than Newton himself would ever have countenanced: practically everything worth explaining, he thought, could be explained in material terms.

Here was a hinge point in the history of physics, a moment when the God of the Bible seemed to recede from the world. The strictest atomists made the case—either unnerving or exhilarating, depending on your perspective—that nature could be self-sustaining. Traditional deities began to look obsolete. Lucretius, a follower of the ancient atomist Epicurus, had once written that "nature is free: she does everything on her own steam.... Divinities play no part."[5] The gods of the Greek pantheon might stand off in some great beyond, watching and laughing. But for Epicureans, the real powers that brought reality into existence were atoms and the forces that moved them.

In the eighteenth century, something like this same theology started to seem plausible again. It was called "Deism": the belief that

God, if he existed, looked very much like Laplace's all-seeing mind. He might stand apart in his eternal instant, beholding the unbroken causal chain that linked every event to every other. But once the machine was set in motion he would not—perhaps he even could not—intervene. Whatever happened was fated; mathematics made it so.

The Man-Machine

According to men like Laplace, whoever knew the positions of particles and the forces acting upon them would know past, present, and future. To this adamantine rule, there could be no exceptions. As above, so below; as outside, so inside: If the new mathematics of Newton described everything, that meant *everything*—including humanity itself.

Understanding nature has always meant understanding man, because man is a part of nature. Ancient and medieval doctors supposed that the bodies of their patients were suffused with four basic fluids—the "humors"—to correspond with the four elements of matter. When the world was made of earth and fire, air and water, then the body was coursing with black and yellow bile, blood and phlegm. Mood swings, character types, even fits of madness and hallucination could be interpreted as signs of poorly proportioned humors in the body, to be rebalanced with adjustments in diet and medicine.[6] When we say anything about the composition of the world, we say it in the same breath about ourselves.

And so, when the world started to look like a machine, then so did man. This idea loomed in many a learned mind as the mechanical philosophy began to take hold. "Life is but a motion of Limbs, the begining whereof is in some principall part within," wrote the political philosopher Thomas Hobbes at the outset of his sweeping

1651 treatise, *Leviathan*. "For what is the Heart, but a Spring; and the Nerves, but so many Strings; and the Joynts, but so many Wheeles, giving motion to the whole Body, such as was intended by the Artificer . . .?"[7]

As the universe machine came into operation, another machine was built to match it part by part, like a golem or a shadow-self: the man-machine. In an unfinished book *On Man*, Descartes had imagined "a statue, or a machine made of earth," containing "all the parts which are required to make it walk, eat, breathe" and perform "all of our functions which can be imagined to proceed from matter alone." He concluded that "when the rational soul is put in this machine, it will have its principal location in the brain," watching coolly over the control room of an automaton made from the dust of the earth.[8]

A mind with enough data could map out more than the orbits of Venus and Mars. It could chart the circulation of the blood, too, and the shiver down the spine at the mention of a cherished name. Sitting with icy dispassion in its fortress of implacable reason, a philosopher's mind could study its own impulse to recoil from the touch of an open flame. It could trace the path of visual stimuli from the outside world, through the eyes and into the brain, where they would form into the face of a lover or a friend. Human experiences, it seemed, were the mere surface of things, and the surface of things—as Democritus knew—was an arbitrary human construct. Underneath it all was matter in motion. The real truth of humanity, like the real truth of the heavens, was to be found in atomic movements that could be understood and controlled.

By the days of Laplace, this vision seemed poised to become reality. During the 1770s and 1780s, Laplace entered into an invigorating partnership with the nobleman Antoine-Laurent de Lavoisier.

Lavoisier came to realize that combustion, the process of burning fuel, involved not the elusive "essence of fire" known as "phlogiston," but rather a gas in the air which Lavoisier named "oxygen."[9] Combining this discovery with Laplace's expert physical reasoning, the two men hit upon an explosive discovery: respiration, the mysterious act by which animals took in air to live, was a species of combustion. The same oxygen which fed the flames of an oven, when taken in controlled quantities into the body of a living man, could help release the energy that set his feet walking and his hands moving. Here was the human machine at work.[10]

Reflecting on Laplace's contributions to chemistry, Lavoisier would write, "Perhaps one day, the precision of the data will be brought to the point that geometry can calculate, in its domain, the phenomena of a given chemical interaction, effectively in the same fashion as it calculates the movements of the heavenly bodies."[11] If so, then humanity itself could become an object of calculation, a machine with inputs and outputs to be measured like the turning of the stars.

All Things New

Under the influence of men like Laplace and Lavoisier, Newtonian mechanics began to look like the one set of rules that could account for absolutely everything, without reference to further action on God's part. Not that Newton's theories won an uncontested victory overnight. Followers of Descartes were resistant to the idea that matter alone could exert such a force as "attraction": minds might attract one another, but bodies could only act on each other by making contact in space. And Newton himself had worried that a totally mechanistic description of things was the first step down the primrose path to atheism. Physics might work beautifully, but it couldn't be the whole picture. Somewhere, there must be a role for God.

Still, even in spite of their author's own misgivings, Newton's equations won victory after victory in France. They had a gifted champion in Laplace. And they had a highly effective propagandist in none other than François-Marie Arouet, better known under his pen name of "Voltaire." After an unseemly brawl with a choleric nobleman in 1726, Voltaire found himself exiled to England. No doubt that experience left him with less than warm feelings toward his mother country. But Britain enchanted him by comparison. And no one seemed to do greater credit to that enlightened nation than Newton, whose powers of understanding were positively godlike.

"If true greatness consists in having received from heaven the advantage of superior genius," wrote Voltaire in his *Letters on the English*, "then a man like Newton—the kind of man that is hardly to be met with in ten centuries—is surely the greatest by far."[12] The cult of Newtonian omniscience was already alight with evangelical fervor in England before the body of the man himself was long in the ground. "Nature and Nature's Laws lay hid in Night," wrote the English poet Alexander Pope on the occasion of Newton's death in 1727: "God said, *Let NEWTON be:* And all was Light!"[13] In *Letters on the English* and *Elements of the Philosophy of Newton*, Voltaire preached this same good news to his readers back in France.

He found ready converts in his fellow *philosophes* of the Enlightenment—philosophers like Denis Diderot and Pierre Louis Maupertuis, whose trust was in the power of human reason above all else. "Newton is our [Christopher] Columbus," wrote Voltaire to Maupertuis. "He has led the way into a new world, and I should like very much to travel there."[14] Like the adventurers who journeyed through the distant western lands of the Americas, Newton had uncovered a great untraveled vista, a virgin universe waiting to be explored by the mind of man. The American plains belonged to whatever imperial conqueror forced himself upon them. But the fresh new cosmos revealed by Newton would yield up its wonders to the power of enlightened understanding.

The birth of a new world, however, is not a painless affair. In France it came with apocalyptic terror and bloodshed, as the old world convulsed in labor pains that doubled as death throes. Britain's colonists in America claimed a country of their own, fighting doggedly against the forces of their ancestral king until the old regime retreated across the sea in 1783. But in France there could be no separate peace, no comfortable distance between the ancien régime and its revolutionary enemies. The fight was for the motherland itself, and the heralds of the future would not be satisfied with anything less than total victory.

By 1792, *The Times* of London could report without exaggeration that "the streets of Paris, strewed with the carcasses of the mangled victims, are become so familiar to the sight, that they are passed by and trod on without any particular notice."[15] The "September massacres" of that year were a foretaste of what would come. The next year began with the decapitation of King Louis XVI and devolved, by the fall, into a fearsome ritual of carnage under the fanatical leadership of Maximilien Robespierre.

The French Revolution began in a bright dawn of hope that the injustices and absurdities of a decadent monarchy could be cleared away. But it evolved, under the influence of utopian zeal, into an insatiable bloodletting. Its leaders were determined not to stop until every trace of the past—every religious superstition, every social hierarchy, every ritual and tradition—was wiped away. "Head after head, and never heads enough / For those that bade them fall," wrote the poet William Wordsworth in recollection of those appalling days.[16] The founders of the French Republic wanted the slate of the world washed clean with the blood of the impure.

The Spirit of the Infinitely Small

What, in the minds of the revolutionaries, was destined to replace the artifice of the old regime? Only the unimpeachable elegance of the natural world, as uncovered by the light of reason. Time itself would begin afresh: once the king lay dead, the National Convention of the French Republic declared in retrospect that Year One of liberty had begun on the Solar Equinox of September 22, 1792. This solemn date would now be known as the first of *Vendémaire*, a new month named for the grape harvest that took place during it.

Other candidates for the beginning of time did present themselves: the revolution had begun in earnest on July 14, 1789, with the storming of the Bastille. But Gilbert Romme, the mathematician charged with resetting the calendar, could not let the petty accidents of human politics dictate the order of the new world. Time and space would now be governed, not by the caprice of mere mortal men, but by the perfect logic of nature, which ruled over a mathematically pristine universe.[17]

"For behold, I create a new heavens and a new earth: the former things shall not be remembered, neither shall they come to mind." In the prophecy of Isaiah, those words belong to God (65:17). But now the human architects of the revolution took divine prerogative upon themselves as they set out to annihilate a wicked past. It seemed almost as if Newton's own "absolute time" and "absolute space" had been uncovered, no longer disfigured by the bracken of arbitrary convention but clear and undefiled, ready to be divided up into a neat and sensible scheme of measurement.

The Republican week did not last for seven days—a mythological holdover from biblical times—but ten. The number ten was orderly, rational, conformable to the mind: everything must fit neatly within it. A new system of weights and measures, the "metric system," replaced the "pounds" and "grains" inherited from the Romans. And the very Earth was organized along a grid of latitude and longitude more precise than in days gone by. The old "Academy of Sciences" was suppressed, and in 1795 (or "Year Three," on the seventh day of "Messidor," the "harvest-giving month"), a new "Bureau of Longitudes" was created. Its task was to "perfect the astronomical tables and the method of longitudes"—to map out the globe and the sky.[18]

It would come as no surprise to anyone that among the officials of this new bureau was one Pierre-Simon Laplace. Being of no great aristocratic birth, Laplace had escaped execution during Robespierre's "Reign of Terror."[19] Now he was granted a place among the creators of a new world. His position was not an especially exalted one, but the victory of his philosophical outlook was entire: the universe was being brought forcibly into order, explained and accounted for with exacting mathematics.

And even if the Republic itself should prove impermanent, the composure of this rationally governed universe was meant to stand forever. As a matter of politics, France was soon governed by a new conqueror, a former student whom Laplace had examined while teaching at the École militaire: Napoleon Bonaparte. Napoleon, himself an aspirant to scientific genius, regarded Laplace with at best a begrudging appreciation. According to one widespread account, he dismissed Laplace from public office with a biting character assessment: Laplace, thought Napoleon, "sought subtleties everywhere, conceived nothing but problems, and finally carried the spirit of the infinitely small into the administration."[20] But whether Napoleon approved or not, this "spirit of the infinitely small" was coming to rule the world.

Whatever government he served—be it monarchic, republican, or imperial—Laplace lived in a cosmos dominated by the small gods. The Republican calendar would lapse into disuse, but the spirit that inspired it would remain alive and well. It was particulate motion that mattered; it was the forces of physics that explained everything. And the old God, of Israel and Christ? When quizzed on the subject by Napoleon, Laplace is said to have replied, "I have no need of that hypothesis."[21] Divine interference was apparently neither welcome nor necessary to explain the unfolding of events. The motion of atoms was all.

The Ingredients of Life

It seemed there was nowhere that these small gods did not penetrate, no place or time where their movements were not at work. The boundaries of human flesh were nothing to them—they *were* human flesh in its simplest components. They were what created both the lungs and the air that filled them, the muscle of the heart and the tiniest drops of the blood that flowed through it.

In 1799, the same year that Napoleon seized power in France, a young English gentleman named Humphry Davy was experimenting across the channel with the effects of what was then called "dephlogisticated nitrous air," now known as nitrous oxide: laughing gas.

Sitting in a sealed chamber and inhaling quart after quart of the stuff, Davy would stumble wide-eyed into the light, giddy with manufactured exhilaration. At one point he emerged so dazed with euphoria that he could only express the transcendence he felt by gasping out a cryptic saying, like some oracle inhaling secret vapors in his dark temple: *Nothing exists but thoughts! The universe is composed of impressions, ideas, pleasures and pains!*[22]

"I feel like the sound of a harp," mused one patient to whom Davy administered the gas. Another test subject, the brilliant but tortured poet Samuel Taylor Coleridge, was hardly less enthusiastic: "I could not avoid, nor felt any wish to avoid, beating the ground with my feet," he wrote.[23] These were not just the ravings of a few dazed junkies. Davy was a pioneer of modern chemistry, a discipline now purged by men like Lavoisier of the mistaken assumptions that had hampered it when it went by the name of "alchemy."

That name, alchemy, came through Arabic from the Greek word *chēmeia*, indicating the mixture of fluids and molten ore to create new and strange substances. And as the search for prime matter developed into the study of atoms, so alchemy was replaced by *chemistry*, a new word from the same Greek root. Chemistry was physics at the atomic level, the study of how the smallest and purest things might flow and blend together.

Davy's object, then, was to separate the core components of "common air" and find out which parts did what. Out of the invisibly subtle mixture of the atmosphere, he and his colleagues aspired to siphon off the raw essentials of life itself. When he called upon his friends and patients to confirm the effects of laughing gas, Davy was touching on a hidden secret that physicians had longed to penetrate for centuries: he was learning to isolate the physical roots of thoughts and feelings.

If "impressions, ideas, pleasures and pains" were the stuff of all human life, and if they could be stimulated with the inhalation of certain gases—well, then they must arise from physical processes in the body. Chemistry was poised to realize every doctor's fondest hope, a hope that has led the way to psychotropic medications like antidepressants and stimulants: the hope that one day, with enough research, the whole complex array of human moods and disorders might come with a list of ingredients.

Those ingredients, of course, would have to be made of atoms. It was not long after Davy's experiments with nitrous oxide that a Quaker named Richard Dalton sat pondering a set of handmade wooden spheres, matching and joining them to one another. Chemists had the tools now to weigh various kinds of air, teasing out gaseous substances and comparing the subtle differences in their composition. Hydrogen was lighter than an equal volume of oxygen, which was lighter again than nitrous oxide. If each substance was composed of elemental atoms, like balls of wood joined together, then the relationship between their weights should be regular: Hydrogen might combine with oxygen or nitrogen, but the weight of each underlying element would stay the same.

And so it was. Between 1807 and 1827, in his *New System of Chemical Philosophy*, Dalton put forward groundbreaking tables in which he compared the relative weights of these elementary particles, the infinitesimally tiny versions of his wooden spheres.[24] They might combine and recombine with one another, but throughout all change and transformation these solid orbs of being—atoms like hydrogen and nitrogen—would endure.

Here were the small gods themselves, laid almost bare—still invisible, but making their existence known on the visible surface of things. Their cumulative presence could be tallied up in the weights and measures of a new study called "stoichiometry"—the measurement of elements. In 1811, an Italian named Amedeo Avogadro discerned that atoms in a gas joined into larger units called molecules, and that "equal volumes of gasses under the same temperature and pressure would contain equal numbers of molecules."[25]

These and innumerable other refinements took years to come into focus. But by the mid-1800s, the underlying principles of the thing were already becoming so solid you could almost touch them, as if the atoms themselves were as tangible as the blocks of metal and wood they composed. It came gradually to look as if the surface world—the world of flesh and blood, flame and stone—was all but a mirage. Real truth was at the level "of the infinitely small": reality's hard and solid bedrock was to be found in the conglomeration of atoms as they coupled and decoupled, drew together and flew apart, to create the fleeting experience of mere perception.

The Making of Man

The world as seen through human eyes—the world of memories and dreams, longings and loves, colors and sounds—was fading into nothingness, a mere mirage. But did that mean humanity itself was expendable? Today, the materialist heirs of the atomist tradition are coming to think so. Psychotropic drugs, however useful they may be for stabilizing certain patients in dire circumstances, have also inspired a mania for chemical solutions to spiritual problems. Humanity is increasingly viewed as a nuisance to be medicated away or eradicated, an ecological stain on the perfection of nature. It will take a new revolution—a spiritual revolution, one whose makings are perhaps already before us—to bring the heart of man back into science. For since the nineteenth century, we have gradually come to feel that humanity has no place in the world.

To the most vigorous and influential thinkers of that century, atoms were not only the source of everything that could be seen and felt. They were also the *means* by which seeing and feeling came about. If atoms made up things like tables and chairs, surely they made up the human eye as well. The machinery of man could now be studied with ever-greater precision, its protocols as regular as the combustion of oxygen or the pull of gravity. The same laws that governed earth and fire must govern breath and bone: it seemed that nature was one vast field of atoms, and whether they made up a hunk of dirt or a human body, they always behaved exactly as the laws of physics dictated. The only questions were which elements were present, and in what quantities.

In 1859, the English naturalist Charles Darwin presented the world with the book *On the Origin of Species by Means of Natural Selection, Or: The Preservation of Favored Races in the Struggle for Life*. Natural historians had been toying for decades with the idea that Earth might be quite a bit older than biblical history would suggest. Darwin was not the first to propose that nature had been grinding out its slow work for millions of years, laboring away with blind regularity to produce the whole panoply of organic life. But Darwin's version of the proposal came with an appealingly clear vision of the whole, at a time when natural philosophy was marshaling its impressive achievements into a comprehensive system of total knowledge.

That word itself—English *knowledge*, German *Wissenschaft*—was coming to replace "natural philosophy" as a descriptor of what men like Darwin were doing. They were thought to be achieving *knowledge* of a uniquely certain kind, based on logical deduction from the hard evidence of the senses. This was not merely one branch of philosophy: this was knowledge itself or, in Latin, *scientia*. This was "science."[26]

Here is what science revealed, thought Darwin: no bolt from the blue was necessary to bring a new species into being. Forms of life were produced out of an organic process called "natural selection." Animals and plants survived if their attributes were well suited to their environment. As English gentlemen bred racing and hunting dogs, culling each new litter to retain those specimens most fit to purpose, so nature ruthlessly executed all but the fittest creatures.

But there was this difference between human and natural selection: "Man selects only for his own good; Nature only for that of the being which she tends."[27] Punishing though she might be, nature was a huntress who spared the best of her children. Every new generation contained some anomalies. Those traits which helped an organism to reproduce would eventually persist.

There remained some mysteries. When Darwin published, he could write frankly that "The laws governing inheritance are quite unknown." They would not stay that way for long, though. In 1866, a Catholic friar named Gregor Mendel proposed that new life-forms were brought into being through "the composition and arrangement of elements which meet in the cell in life-giving union."[28] Though gene theory in its fully developed form would have to wait until the next century, the first germs of it were taking shape. It gradually became clear that atoms and the molecules they formed were at work even in the creation of new life.

Humanity itself, it came to seem, had arisen from the churning of atoms. Colliding and combining by the iron laws of nature, these implacable beings created the very minds that now presumed to study them. This all-consuming explanation, this faith in the small gods, reached its height as the nineteenth century drew to a close. The world was a machine—and so was man, locked into the path laid out for him by the regular movement of atoms through space.

The Final Trap

"There is grandeur in this view of life, with its several powers, having been originally breathed into a few forms or into one," wrote Darwin at the close of *On the Origin of Species*. "Whilst this planet has gone cycling on according to the fixed law of gravity, from so simple a beginning endless forms most beautiful and most wonderful have been, and are being, evolved." An attractive new myth of man's ascent out of primordial chaos. But was humanity really being raised to dominion over nature—or simply absorbed back into it?

The knowledge of physical law was supposed to give man power over the world around him, and in a certain sense it was doing so. The poorest steerage passenger on a steamboat could benefit from a mastery of the waves that the boldest magicians of the Renaissance had only dreamed of commanding. At the same time, though, even the most sparkling insights of the sharpest minds were being reduced to mere glimmers on the surface of an atomic sea, bright sparks thrown off from the inexorable collision of particles as they crashed together.

In the 1840s German scientists, working on a theory put forward by the psychiatrist Wilhelm Griesinger, began picking apart the skulls of deceased maniacs to uncover the source of their illness in the gore and slop of their broken brains.[29] But it seemed that even healthy thoughts might bubble up like so much froth from cerebral matter, just as regularly as in any other physical process. In 1863, the Russian physiologist Ivan Sechenov put the matter definitively: "The brain is a mechanism which, if brought into action by a certain cause, ultimately produces a series of external phenomena." The "excitation of the sensory nerves" produced "excitation of the spinal center linking the sensory nerves with the motor nerves," followed in tight succession by "muscular movement."[30]

From the twitch of nerves to the spasm of muscle, from brain to body by way of the watching mind, human life was a mechanism like any other. It started to seem to the worshippers of the small gods as if reflection and choice were fictions, representing the patient's dim awareness of a biological machinery that only happened to toss up ideas as by-products. In the new world of atomic motion, a man's thoughts and feelings were thrown up against the screen of consciousness by the moving parts of the body and brain, just as qualities like color and shape emanated as trivial aftereffects from the atoms of the outside world.

"Water, in flowing, hollows out for itself a channel," wrote the French psychologist Léon Dumont in 1876: "Just so, the impressions of outer objects shape more and more appropriate paths for themselves in the nervous system."[31] The American philosopher William James quoted Dumont at length in his massively influential *Principles of Psychology*, which hit shelves in 1890.[32] By 1915 Sigmund Freud, father of psychoanalysis, could toy with the idea that "the individual is a temporary and transient appendage to the quasi-immortal germ plasm."[33]

The mind of man itself was to be dissected, his ideas and emotions reduced like every other natural phenomenon to the thoughtless interaction of fluids and meat. Here was the total victory of the small gods. Here was humanity forced to kneel at the altar of powers it had tried to control. "The intellect and the mind are objects for scientific research in exactly the same way as nonhuman things," Freud pronounced at the close of his career: neither piety nor pity would keep the scientist from subjecting all things to the laws of matter in motion.[34]

But if even the mind was to be an object of physical study, who would do the studying? If thoughts and experiences were to be analyzed into particulate matter, what about the thoughts of the analyst himself? Those, too, must be products of brain chemistry—and if the chemistry obeys mathematical laws, then so do the thoughts. If thinking itself is the result of a predetermined mechanical process, all we can really do is watch as the chain of causation unfolds. If the world is subject to the motion of atoms, then the mind of man is like Laplace's demon or the distant God of Deism: trapped and immobile, helpless to do anything but look on as the laws of physics perform their irreversible work. If the human body is really a machine, the mind must simply sit and observe as the logic of biology plays out.

Laplace's world of totally determined motion once glittered with the promise of perfect knowledge. But its shine was that of a world frozen in ice, forever still and unmoving. In it, even the human mind became a product of mechanical motion, as all of history and thought was stuck in place. Nothing could be otherwise than it was: past, present, and future were seemingly locked in step with the ironclad rules of computation.

Gradually, this kind of thinking subjected everything—first the wheeling of the planets, then the burning of flame, then finally the breath of life in the lungs of man and the play of thoughts in his mind—to one set of frozen laws. Like a king in a fairytale castle, sitting high in his citadel while the realms around are frozen by the chill of an eternal winter, mankind sat and watched as the curse drew nearer and nearer—as first his earthly home, then his body, then the mind itself that did the watching, fell beneath the spell of the frozen world.

CHAPTER 5

A Strange Kind of Chemical Change

And I'll be dreamin' of the next time we can go
Into another serotonin overflow.
> —John Mayer, "Love on the Weekend"

Gradually, the world was going dark. Those who reduced humanity to its material origins found themselves hemmed into a closed system, trapped in the increasingly mechanistic jaws of nature's logic. A species that once looked to the stars for its origin and its destiny had now to scramble to find new purpose in a frozen world stripped bare of the spirit. A desperate task, and ultimately a futile one—but not for lack of trying. The lords of the earth and the captains of commerce needed to build a new mythology for themselves, a story that could invest the future with hope and meaning. The story they came up with is still with us today. It goes like this:

Man began among the beasts. But he will not stay there. The creature that once huddled naked under the open air, foraging in the dirt for food, jittering and cringing at every unknown snarl in the grim night—this same pitiable wretch will one day master all of nature. Wielding the awesome powers of his new science, he will rest in proud serenity amid the glories of a domesticated world.

This was supposed to be the promise of the twentieth century as it dawned. At that point, steam trains were already charging across mile after mile of open country like muscled stallions, snorting out clouds of smoke as if in derision at the once-imposing size of the terrain. The sky was next: in 1903, a machine that its inventors simply called "the Flyer" would lift off the sands of North Carolina for twelve immortal seconds. As Wilbur and Orville Wright pieced their airplane together, inventors across the world were competing to send a "wireless telegraph," later to be known as a "radio signal," over miles of space in an instant.

Movie cameras and record players, tea bags and air conditioning, vacuum cleaners and steering wheels: these sublime instruments of power and comfort would now sit casually in the same simian hands that once groped for rocks and lumber in the dark. Mankind could now go riding on the currents of the wind like some favored demigod; his voice could throb in the air like a command from Olympus. Soon, nature's child would become her lord; technology would grant him that dominion. Or so the titans of industry came to believe.

The new science didn't only wield power over time and space. It also captivated the human imagination. Year after year, amid the high drama of business deals and international competition, an entirely new vision of humanity itself was taking shape. To the most enthusiastic advocates of this new myth, it seemed that mankind was becoming not just nature's master but her product—an object of her exclusive craftsmanship. She had supposedly formed him the same way she formed everything else: blindly, as dictated in the script set out for her by the laws of physics. It was a matter of indifference whether that script produced a mollusk or a clod of dirt.

But now it had produced this strange ape, whose mind had been honed to an ever-increasing effectiveness by unsparing generations of trial and error. This poor monstrosity came trailing clouds of illusion that he called by names like "love" and "virtue," as if the advanced development of his brain had overheated it and plagued him with uncouth fantasies.

It was for him to sort all this out, of course. The luck of the evolutionary draw had equipped humanity with an extraordinary calculating machine. Its processing power outstripped anything previously in circulation. But the by-product of this unplanned experiment was a whole weight of longing and remorse, consciousness and conscience. If man was brilliant, he was also deformed: the healthy animal functioning of his body was stunted by the lopsided development of his febrile nervous system. Now he must stumble around upright, dreaming bad dreams.

Human consciousness was nothing more than a by-product of naturally occurring physical phenomena: gradually, this revolutionary idea triumphed so completely that it could almost be taken for granted. In 1956, for instance, the debonair jazz icon Frank Sinatra could croon philosophically about "How little we know / How much to discover / What chemical forces flow / From lover to lover." Scientific knowledge was still limited enough that the better part of wisdom was to toss it aside with a gleeful shrug in the frenzy of romantic abandon: "So long as you kiss me / And the world around us shatters / How little it matters / How little we know." At the same time, though, it was understood almost as a matter of course that even the lover's bedchamber would not long remain off-limits to laboratory analysis. The day was coming when all the sources of human passion would be catalogued, and it was widely assumed they would be chemical.[1]

By 1959, Nat King Cole could expect everyone to take it in stride when he sang that "Something happens to me / Every time I feel that you are near: / A strange kind of chemical change goes rushing through me."[2] Those words sound unremarkable, even banal, in the twenty-first century. But before the mid-1900s, no casual listener would have nodded along to hear the thrill of fresh romantic affection described as "a strange kind of chemical change." Those words represent a new assumption taking hold, a chic new theory according to which "chemical" could be used almost as a synonym for "emotional."

If the glimmer and exaltation of human feeling is actually just the aftershock of biological energy transfer, then emotional change *is* chemical change. Talking about one can become shorthand for talking about the other. Today, that shorthand is so commonplace we don't even notice it. But in the age of steam engines and first flight, it was still taking hold. In fact, that era—the era of the atom and the machine—is what gave us our mechanical and chemical idea of ourselves. It is like a genetic inheritance that we cannot seem to shake.

Shameful Origins

No one saw the coming change in mankind's self-understanding with starker clarity than Friedrich Nietzsche. A pastor's son from the German state of Prussia, Nietzsche was born in 1844 into what should have been auspicious times. King Friedrich Wilhelm III had declared decades before that Prussia's two major denominations, the Lutherans and the Calvinists, should lay down their disputes and merge in Christian brotherhood. "May that promised point in time not be far off," murmured the king, "when all, under one common Shepherd, will form one flock in one faith, in one love and in one hope!"

But underneath his dreamy smile, the enlightened monarch betrayed more than a touch of haughty indifference. The churches, sighed King Friedrich, "are divided merely by certain external differences." Participants in the dispute might consider those differences theologically crucial, but from the divine perspective of the lofty throne they were mere bickering and distraction.[3] Let matters like the predestination of souls be set at rest—they would never be decided anyway. Get on with the business of serving the German state.[4]

Nietzsche's painful talent was to see straight through this kind of bureaucratic double-speak. His unsparing eye could pierce even the most delicate fictions. Be it a blessing or a curse, he was willing and able to debunk every pretty lie and self-delusion that kept modern men in courteous harmony with one another.

So unreflecting patriots might tell a heartwarming story that humanity was moving beyond religious hostility. But Nietzsche could see that disputes over eternity and salvation were becoming less violent only because they were becoming less important. Behind the comfortable façade of high society and the smooth efficiency of the Prussian state, there yawned a black chasm of emptiness—a nihilism too shattering for lesser men to name. The nations of Europe were learning to focus on material prosperity and earthly peace because they cared less and less about anything else.

Most infuriating of all, people hardly even seemed to notice the transformation unfolding. They went on chattering about niceties like "justice" and "virtue," as if those words referred to absolutes in some shining realm of eternal glory. But in the same breath, educated men were already talking as if no such realm existed, as if the cold light of science had dispelled the phantoms of the supernatural world once and for all. Animal evolution and material law explained everything; the world was made of nothing but fragile sinew and cold dust. Where was God in all that? What right had evolutionists to speak of realms beyond? Disgusted with the pretentious abstractions of his fellow academics, impatient with the theologians whose verbal tricks he knew so well, this tortured prodigy set his formidable mind to exposing the rotten foundations of conventional morality.

"There was a time when one looked to feel man's greatness by indicating his divine *origin*: this way is forbidden now, for at its entryway stands the ape," wrote Nietzsche in 1881. Man's *pudenda origo*, his "shameful origin" in mere nature, revealed him as a pitiable thing of flesh. Whoever grasped this truth must think "with contempt of the warm, comfortable, misty world in which the healthy man wanders unthinkingly."

The blissful ignorance of a middling intellect is to think two thoughts at once: science can reduce the world to the motion of physical objects, but civilized men can go on treating one another with charity and brotherhood as if they had a reason to. Nietzsche kept insisting that if the outside world is simply a product of physical forces, then so is the supposed "inner life" of moral vision and aspiration. "We have spent so much effort learning that external things are not as they appear to us to be—well then! The case is the same with the inner world!" Nietzsche read this message into the supposedly wonderous new discoveries of science: deep beneath the surface of our nobler desires and the stirrings of our heart, deep beneath even our dreams, lies the truth that man is a body, destined for the grave.[5]

The Madman and the Craftsman

Nietzsche hinted at all this in his 1881 book, *Morgenröte* (*The Break of Day*). The dawn he saw on the horizon was dim, and the world it would reveal was still invisible to those who walked about in darkness. To decent people—well-adjusted, productive, respectable people—a man like Nietzsche sounded like a madman, gibbering incomprehensibly that morality had decayed into a mere sham. One year later, in the first edition of *The Gay Science*, Nietzsche put his most famous observation into the mouth of just such a lunatic, screaming in the streets at befuddled passersby: "God is dead—and we have killed him!"[6]

In *The Break of Day* Nietzsche had asked, "Do you understand why it had to be madness that did this?" He himself would succumb to a madness of his own, dissolving at the end of his life into a wreck of nervous desperation and thwarted hope. But the cold daylight that he saw coming did indeed break, and his talk of a soulless universe did not sound insane for long. Though he was ahead of his time, Nietzsche was not the only one who could see mankind transforming into something altogether unexpected. The silhouette of this new humanity could already be discerned in the dense and impassioned writings of his fellow Prussian, Karl Marx.

As a young man elbowing his way into the contentious philosophical circles of Berlin, Marx had wrestled with the tangled relationship between history and ideas. What was this creature called man, and what did he have to say about his fate? Working at first under the shadow of senior eminences like Georg Wilhelm Friedrich Hegel and Ludwig Feuerbach, Marx arrived at last at a bold proposal of his own: abstract thought, however beautifully expressed, was a by-product of material events. "Men can be distinguished from animals by consciousness, by religion or anything else you please," he wrote with his collaborator Friedrich Engels in the manuscripts known as *The German Ideology.* "They themselves begin to distinguish themselves from animals as soon as they begin to *produce* their means of subsistence, a step which is conditioned by their physical organization."[7]

For Marx, history began when one small corner of nature became self-aware. Humanity was that wrinkle in the fabric of things, the hinge in time where the world folded in on itself and started to direct its own development. As soon as he could recognize his own needs and set out to fulfill them, man rose from the clay that formed him and began to shape it in turn. Engels added later that this process was teetering on the verge of total fulfillment, leading toward a moment when creation would come "under the dominion and control of man, who for the first time becomes the real, conscious lord of nature. . . . Only from that time will man himself, with full consciousness, make his own history."[8] This new man would take up his plough and his anvil to remake the universe in the image of *homo faber*—the craftsman of the world.

"A spider performs operations that resemble those of a weaver, and a bee puts plenty of architects to shame in the construction of her cells," reasoned Marx. "But what distinguishes the worst architect from the best of bees is this: the architect raises his structure in imagination before he erects it in reality."[9] Other animals might clutch at food by instinct, with no more forethought than a raindrop falling from a cloud, and for the same reason: because the brute forces of nature made it necessary. But now those same brute forces dictated that human beings should take matters into their own hands and labor to reshape the universe according to their own needs.

The truth about man, then, was not painted on the ceilings of cathedrals or laid bare in strains of poetry. It was inscribed into the shape of his tools: "relics of bygone instruments of labor possess the same importance for investigating extinct economic forms of society, as fossil bones possess for the determination of extinct animal species," wrote Marx in *Das Kapital*.[10] Technology and science, the products and procedures of physical satisfaction, explained the nature of humanity. Morality was a parlor game unless it referred to this fundamental reality, this basic imperative to muscle things into conformity with man's needs. Discarded tools lay buried in the earth like the skeletons of fallen civilizations, revealing the central secret of life: man is man because he builds. Nothing could be higher or truer than that.

Rule by Machine

No prophet is honored in his homeland. Marx spent his whole life dreaming of a day when the workingman would realize his birthright as creator of the world, dispelling the self-serving fictions woven by decadent aristocrats to keep their servants docile. But it was not in his native Germany that this prediction was fulfilled: it was in Russia, more than thirty years after Marx's death, that the revolution came. In October 1917, in the second insurrection of that tumultuous year, Vladimir Lenin and his Bolshevik Party surged to power over the wreckage of the Tsarist empire and the flimsy provisional government that stood in its place.

Years of slaughter and intrigue were still to come as the leaders of the new socialist vanguard jockeyed for control. But there was one thing about which practically everyone in the new regime agreed: the high-minded pretensions of religion and tradition had been exposed as instruments of oppression, used to justify the violent exploitation of the common man. "As for us, we were never concerned with the Kantian-priestly and vegetarian-Quaker prattle about the 'sanctity of human life,'" seethed the Marxist firebrand Leon Trotsky.[11] Ideas like "God" and "soul" were disgusting relics of a guilty past. They must all be torn apart and burned.

The revolutionary year 1917 saw the founding of Proletkult, the "Proletarian Cultural and Educational Organization." One member at the inaugural meeting, as described by the historian Richard Stites, declared "that all culture of the past might be called bourgeois, that within it—except for natural science and technical skills ... there was nothing worthy of life."[12] Like threadbare drapery torn from the walls of a fallen palace, the old legends of the aristocracy must fall now to be trampled in the mud. The truth of things—the rational, mechanical, scientific truth—must be unveiled.

Man's origins could be seen plainly in the bodies of the other animals, whenever a scientist peered into the split belly of a dissected frog.[13] The mute witness of pulsing guts could testify that human beings, too, were nothing more than a tangle of intestines and nerve endings. And if man had come from the forge of nature, honed over eons to survive, then he must now use his tools to continue that process. A new dream of history's end point blazed ahead, eclipsing the old stories of Armageddon and resurrection: not in some faraway heaven but here, on earth, man would take ownership over his evolution. His tools would become part of him, his flesh fusing with steel in a merciless ecstasy of perfection.

"Future society will be managed by special 'production complexes' in which the will of machinism and the force of human consciousness will be welded together unbreakably," wrote the revolutionary poet Aleksei Gastev in 1918.[14] Man, Gastev supposed, would soon dissolve into a sea of interchangeable parts, his very heartbeat regulated by the supreme will of steel and numbers. He would then be rid at last of those tormenting specters called "love" and "anguish," those wisps of nonexistence that had clouded about his sight since he climbed from the primordial sea. He would stand reborn, "soulless and devoid of personality, emotion, and lyricism, no longer expressing himself through screams of pain or joyful laughter, but rather through a manometer or taxometer. Mass engineering will make man a social automation."[15]

This frigid prophecy was hardly less bleak than the reality that would materialize under the rule of Lenin's successor, Joseph Stalin. At the command of this unyielding zealot, justice and mercy alike were discarded as cheap sacrifices on the twin altars of efficiency and compliance. The new law decreed that all manufacture, all wealth, and all allegiance should belong to the Party alone. Before the tribunals of this dictatorship, the worth of a human life was nothing except a measure of its contribution to the Soviet machinery.

Nikolai Krylenko, who served as an officer of justice under both Lenin and Stalin, declared baldly that when a defendant stood before the Soviet courts, "only one method of evaluating him is to be applied: evaluation from the point of view of class expediency." Looking back on the frenzy of imprisonment and paranoia that defined those years, the dissident Aleksandr Solzhenitsyn recalled: "people lived and breathed and then suddenly found out that their existence was *inexpedient*."[16]

This heartless machine-state, fueled with the blood of millions upon "inexpedient" millions, did eventually collapse. Krylenko himself was a victim of Stalin's "Great Purge" in the 1930s. The Soviet Union fell six decades later, brought down by American pressure from without and the capitulation of the Soviet government from within. Neither the world at large, nor Russian leaders themselves, could ignore any longer the economic failure of the regime, the unraveling of its empire, and the growing restiveness of its beleaguered and suppressed people. But for all that Stalin lies dead in the ground, and for all that the Berlin Wall lies in rubble, some remnant of the Soviet spirit still haunts the world. The logic of Stalin's regime was more powerful and enduring than any one government. It arose directly from the intoxicating promise of an industrial age: the promise of rule by machine.

The Metal Mind

While Stalin clutched furiously at autocratic control, an altogether different kind of power was taking shape far to the west, in England. Its chief exponent was Alan Turing, who made his name during the Second World War as he and his countrymen raced to outgun and outsmart the Nazi military. Courage and determination were of the essence, as they had always been in battle. But everyone involved—from Joseph Stalin to Adolf Hitler to England's own Winston Churchill—knew that this war would not be won by soldiers alone. More certainly than ever before, victory belonged to the nation whose scientists could master the most ingenious technology.

It was in this respect that Turing excelled. His most urgent contribution to the war effort was to help perfect a device called the bombe, which could decode German messages by winnowing down the possible arrangements of letters that might have been used to scramble the text. The bombe would sift methodically through huge reams of possibilities. All but a very few would generate a logical contradiction, leaving the rest at a manageable number for humans to sort through and crack the code.

The possibilities inherent in this kind of thinking were far from exhausted when the war ended. Turing was on the verge of designing a machine that could proceed automatically through any possible calculation: a machine that could compute. Before the war he had described an "a-machine" ("a" for "automatic"), which operated exclusively by detecting a series of prompts written on paper and responding with a small number of scripted procedures. "It is my contention that these operations include all those which are used in the computation of a number," he wrote. This device, soon to be known as the "Turing machine," could move methodically from step to step, writing and reading only the characters 1 and 0 plus the placeholders ə and x. These limited methods, Turing showed, could perform any numerical calculation required.[17]

A simple enough claim on its face. But in application, it was a thing of limitless potential. "The memory capacity of a machine more than anything else determines the complexity of its possible behavior," Turing wrote after the war. With enough storage and enough time to execute on instructions, a computer could chew through vastly more calculations than even the greatest human mind, with vanishingly small error.[18]

Turing's predecessors had dreamed wistfully of automated calculators, and some had even drawn up plans to make one—in the nineteenth century, Charles Babbage and his dedicated interpreter, Lady Ada Lovelace, had outlined a theoretical model for computing machinery.[19] But Turing was poised to make the far-fetched imaginings of bygone tinkerers into hard and fast reality. Developed to their full potential, Turing machines could perform feats of calculation and organization beyond the wildest Soviet fantasy. Perhaps, given time, they could even think.

For if the brain was simply a crude machine, why couldn't technology be used to improve nature's haphazard constructions? Turing was determined to show that the mind of man was essentially interchangeable with a sufficiently advanced computer. His creations were processors of discrete information, and "brains very nearly fall into this class . . . they could have been made to fall genuinely into it without any change in their essential properties."[20]

In 1950, outlining what would come to be known as the "Turing test," he argued that a machine could "think" if it could convincingly pose as a human. Isolated in a dark room, an examiner could interrogate such a machine alongside a human to sniff out differences in each one's response. Once the stilted jerks and glitches of its protocols had been smoothed away, the machine would have every right to claim the mantle of conscious thought: at that point, both participants in the test would merely be competing sets of advanced circuitry, turning inputs into outputs.[21] And if the machine shared all the important attributes of the man, perhaps the man could take on the superior capabilities of the machine. Humanity would then be ready for an upgrade.

Building God

For the machines, at least, the upgrades came thick and fast. Computer technology has soared to staggering heights of sophistication and complexity, just as Turing hoped it would. Once the internet linked computer to computer at unheard-of speeds, these imitation minds began to seem as if they could do almost anything.

Soon the average household found itself occupied by a wakeful robotic sentinel, a box of unknown mathematical procedures hiding behind a colorful screen. "It's an alien life form," said the rock star David Bowie of the internet in 1999.[22] The more indispensable they became, the more inscrutable computers were to their average user, the elaborate architecture of their programming hidden and compressed within the recesses of their sleek bodies.

It seemed they could do everything mankind had dreamed of doing all these years: they could share information across great stretches of space and commune with one another in an instant. The simplest of them could assimilate great troves of data that would dumbfound the human mind, sorting through it with the impartial rigor of mathematical logic. Computers assessed the world without the distorting mediation of passion or desire, just as the heralds of the scientific age had longed to do. Their only imperatives were those of necessity.

In the wake of these epoch-making developments, the old longing for a sublime fusion between man and machine has today taken on a kind of messianic fervor. "Our species has been rapidly acquiring superpowers in the form of unprecedented levels of control over our bodies, brains, built environments and ecosystems," writes Elise

Bohan of Oxford University's Future of Humanity Institute. Just as Marx predicted, humanity has risen up to reshape its environment in the image of its machines—and now, say the prophets of the cyborg future, man himself must become the object of his own unsparing optimization. "We're loath to admit it, but the world is not set up for ape-brained meatsacks any more," declares Bohan. "We've seeded an extra-human layer of thought. . . . The next step is to bring it to life."[23] These are no longer the fever dreams of a few tech enthusiasts or Soviet fanatics.

Transhumanism—the promise of a new generation with bodies and minds enhanced by machine technology—has become a millenarian aspiration among the magnates of the world. In 2022, President Joe Biden's White House announced via executive order its intention to invest in technology that will make its wielders "able to write circuitry for cells and predictably program biology in the same way in which we write software and program computers" so as to "unlock the power of biological data."[24] If man is a prototype computer, constructed from the raw materials of an uncaring nature, then his destiny is to update himself into eternity. Stretching out his hands to touch the blessed diodes of his most cherished creations, he will slot himself at last into the coiling network of information and electricity that is already snaking through the air and earth all around him.

As artificial intelligence programs have learned to mimic human language and creative output to an eerie degree of specificity, these kinds of predictions have only become more dramatic and extreme. One acolyte foretold the coming of programs that can store and process infinite knowledge: "WE ARE CREATING GOD WHAT DON'T YOU UNDERSTAND." Elon Musk, one of the world's best-known advocates of the man-machine merger, replied: "Exactly."[25]

Breaking the Circuit

This dream is fast becoming a nightmare. For all the promises of liberation and triumph, our bodies and brains have not responded well to our efforts at treating them like first-generation meat computers. No matter how finely we calibrate our hormone levels with pills and injections, our existential despair and unease only deepens.[26] Swapping out genitals and body parts like plug-ins and accessories has not yielded promising results. Transgender surgeries, which one enthusiast hoped would break a "lucky few free from the horrid curse of being human," have instead worsened that curse unspeakably, leaving patients to suffer through infection and daily agony.[27] And as artificial intelligence takes on levels of complexity that even its creators can't predict or control, dire misgivings about its destructive potential have started to take hold. Because if we are simply suboptimal versions of our own machines, why should they meld with us at all? Why shouldn't they enslave us, or scrap us for parts, or simply do away with us?

If the long-awaited advent of the cyborg world is upon us, we will be forced to consider whether this is really what we want—and what other choice we may have. Struggling like butterflies pinned to a circuit board, the skeptics among us are looking for some way out of a world made cruelly flawless by our own inventions. Eliezer Yudkowsky, a pioneer in the development of artificial intelligence, described his fears for what the technology might become: without the right stopgaps, he wrote, we will end up with a situation in which "the AI does not love you, nor does it hate you, and you are made of atoms it can use for something else."[28] But this grim future is far from inevitable. It only feels that way to some of us because of ideas we inherited from men like Laplace and Marx, who taught us to think of the world as a machine with a ghost inside called humanity—a ghost destined to evaporate, leaving behind only the ceaseless grinding of gears.

But these ideas are hopelessly outdated. Our philosophical assumptions *about* science are hundreds of years old. They have not kept pace with scientific knowledge—indeed, not even many AI visionaries (or dystopians) seem to understand this. We still think in terms of matter in motion, of atoms and energy, tiny chunks of material colliding, attracting, and repelling in an endless cosmic flow. And so of course we cannot see what should separate or protect us from our robot children.

At the beginning of the twentieth century, the atheist philosopher Bertrand Russell wrote that "mathematics, rightly viewed, possesses not only truth, but supreme beauty, a beauty cold and austere, like that of sculpture, without appeal to any part of our weaker nature, without the gorgeous trappings of painting or music, yet sublimely pure, and capable of a stern perfection such as only the greatest art can show."[29] More than a hundred years later, this attitude persists in the back of our minds, like programming we can't control: numbers are the most sublime truth, and matter is the most real thing.

But deep as these convictions have sunk into our bones, they have not proved correct. All these years, the very science that brought us the steam engine and the digital computer has been teaching us more about both matter and mind than has been incorporated into our popular understanding. The world is far more than a machine; the brain is something different altogether from a computer. These facts have been evident in the scientific record itself for decades. A rich white light has been streaming into what we thought was the metal prison of a purely material world—only we haven't yet raised our eyes to see it.

The way out of our anxiety about humanity's future is not to hearken back to a prescientific age, but to look ahead at what science is actually revealing to us, which is not the well-established mythology of matter. The discoveries of physics point to a more humanizing vision of the world—if we can take stock of all that they imply. We know more than we have yet realized, and our next enlightenment will be far more profound than the last.

CHAPTER 6

Piercing the Veil

"But do I see you as you really are?" he asked.
"Only Maleldil sees any creature as it really is," said Mars.
"How do you see one another?" asked Ransom.
"There are no holding places in your mind for an answer to that."

—C. S. Lewis, *Perelandra*, chapter 16

A toms came up from the depths and formed the universe: that is the story we have come to tell ourselves. It is a story of matter moving blindly through space, its course determined by unfeeling numbers. But that is not the whole of things. It is not even the whole of what science has to say. Another story has also been unfolding—a story not of rigid material mathematics but of life and relationship. It began with light.

Back in the glory days of England's empire, a lowborn prodigy named Michael Faraday made his way from a dingy borough near the Thames to a place of honor in the heart of the capital city. The son of a blacksmith, he had been put to work early, apprenticed to a bookbinder at fourteen. Any sort of advanced schooling was practically out of the question a poor boy like him. It didn't matter: Faraday wrung a homemade education out of every book he could find. The Westminster offices where he worked were a twenty-minute walk from the Royal Institution, a research organization where he could sit in on public lectures by none other than the eminent chemist Humphry Davy. By 1813, Faraday had composed his own three-hundred-page book based on what he heard from Davy. By the 1840s, he was giving lectures of his own.

Above all, he never outgrew his breathless fascination with electricity. In his hands, metal wires became sources of astonishing wonders. Before an admiring London public, he demonstrated that when a magnet passed through the middle of a wire coil, it sent a shock spiraling along the wire. As the magnetic forces pushed their way through space, an answering charge of electricity curled around their path. An electric current flowing in a circuit likewise found its counterpart in a magnetic charge looping around the tracks of the electricity—the interplay of the two could send a motor spinning, or generate electric power from the simple rotation of a metal disc between two magnetic poles.

Among his colleagues, Faraday confessed to certain unorthodox suspicions about the nature of the forces he was toying with. He filled his notebooks with sketches of them, thin black lines curving in delicate filigree through the air. But how did they move? In a strictly mechanical world, there could only be one answer: for the push of magnetic force to be felt, something must be *doing* the pushing. It followed that the force must shimmer through some delicate substance, a fluid or flame so subtle that it evaded touch and sight. Motion could never be communicated through empty space—there must be matter of some kind that was moving, an "ether" whose unseen regions could flutter against one another faster than thought.

But no experiment could find this ether or tally its effects; no measuring device could catch even a whisper of its motion. And if it could never be detected by any human sense, however enhanced, then by definition it could hardly be said to exist as a physical thing at all. Faraday had another explanation, a "view of the nature of matter which considers its ultimate atoms as centres of force, and not as so many little bodies surrounded by forces."[1] Instead of an empty container called "space," occupied by atoms of matter, Faraday imagined a sea of pure force washing outward to the ends of the universe, varying in strength at every point.

Wherever they were closely packed together, thought Faraday, these forces could be experienced as solid objects, pushing against the touch. In their densest centers of concentration, they amounted to atoms—they were what observers experienced as small, indivisible chunks of matter. Looser regions of force were what we typically refer to as "empty space," because we can walk through them with only the tug of things like gravity and air resistance to contend with. But everywhere there were forces, sinuous and ever-shifting, to define the contours of reality. In fact, Faraday argued, force *was* reality: "I do not perceive in any part of space . . . anything but forces and the lines in which they are exerted."[2]

If this was right, then a wave of force would need no physical medium to travel through—no fluid like water or particles of air to carry the billows of its movement from one point to another. The forces themselves were the medium, the rocking cradle where the world of solid things was held. A ripple in the intensity or direction of a force could pass seamlessly from one region to the next, like the invisible flow of energy that could glide through a wire around the shifting poles of a magnet. If it shot across the lines of force that snaked through the space between tangible objects, a wave like that could bring energy darting from the stars above to glimmer undiminished on the distant earth.

"The view which I am so bold as to put forth," wrote Faraday, "considers . . . radiation a high species of vibration in the lines of force which are known to connect particles and also masses of matter together."[3] If there was no such thing as empty space—if every corner of the galaxy was charged with some measure of force—then a throb in that great ocean would be enough to carry the rain of warmth and power that had poured down onto the earth since before human eyes ever opened. No chariot was needed for the sun god any more, no crest of air for his steeds to waft along: a wave in the field of forces could amount to a ray of light.

Ethereal Visions

For all his triumphant achievements, Faraday lacked the training needed to make these claims concrete. That honor would go to a

Scotsman, James Clerk Maxwell, who summarized what had been discovered about electricity and magnetism in four sets of elegant mathematical functions. A shift in one form of energy always created a "transverse" shift in the other, meaning that electricity and magnetism fluctuated together at right angles to one another. Most astoundingly of all, chains of those fluctuations moved through a vacuum at roughly 300,000 kilometers per second—within striking distance of the best available measurements for the speed of light. "We can scarcely avoid the conclusion," wrote Maxwell in 1862, "that *light consists in the transverse undulations of the same medium which is the cause of electric and magnetic phenomena.*"[4]

Here was Faraday vindicated, in a matter of decades: light was a ripple of electric and magnetic energy, cascading together at breathtaking speeds across huge distances. What Maxwell did would become known as the "Second Great Unification." The first belonged to Newton, who drew together every kind of movement into one delicate system—a grid of space whose laws extended from the core of the earth to the endless outskirts of the cosmos. Now Maxwell had gathered together the patterns of light, electricity, and magnetism, bringing the frenetic twitches of their waves into smooth alignment. Two masterworks of the human mind—two tapestries of logic draped expansively over the contours of the whole universe.

Except they did not line up.

Newton's picture of the world was matter in motion—a machine built up from the tiniest atoms to the planets with their stable rhythms. Maxwell's was energy in flow, an unbroken sea of continuous forces billowing through the air. Between these two sets of ideas there was an awkward juncture, like a mismatch between two bolts of fabric that could not be sewn together. In attempting to transition between Newton's solid bodies and Maxwell's energy fields, the mind sputtered and glitched: what physical mechanics were at work when light flickered through space?

Maxwell still pictured light flowing through the ether, that elusive breath of almost-nothingness that had wafted through the scientific imagination for centuries. But this silent specter seemed utterly implausible as the host for a wave that could travel at light's blistering speeds. The faster a wave moves, the more rigid its medium must be—but the ether was supposed to be so slick and yielding that it couldn't even be felt. If light waves traveled through a physical substance, the way ocean waves roll through water, then that substance would have to perform physically impossible feats. It would have to be lighter than a breath and stiffer than steel, undetectable but omnipresent—like some spirit of the air.[5]

Maxwell's electromagnetism was making impossible demands on the Newtonian physics that was supposed to govern objects, straining the older system to its breaking point. What if Maxwell's light could not possibly shine through Newton's space?

The Miracle Worker

At the turn of the twentieth century, an obscure young man sat in a Swiss patent office, worrying over these problems like a dog with a bone. Looking back on his youth, Albert Einstein remembered being tormented at the prospect of squandering his future: "the nothingness of the hopes and strivings which chase most men relentlessly through life came to my consciousness with considerable vitality." He had shown brilliant promise as a boy, but the world had judged him a layabout—too undisciplined to amount to much. Now here he sat behind a desk, rubber-stamping other men's inventions as his own shimmering potential seemed to fade away hour by hour. And all the while there were tremors shivering through the very foundations of the cosmos.

"All physicists of the last century saw in classical mechanics a firm and final foundation for all physics, yes, indeed, for all natural science," he later explained. But this "dogmatic faith" in atomic motion, this worship of the small gods, simply could not explain the behavior of light.[6] Scrounging together the books and journals at his disposal, working almost alone, Einstein proceeded to take physics down to the studs and rebuild it. In 1905 he published four papers, each of which would help redraft the map of the universe. It would come to be called his *annus mirabilis*: his miracle year.

No merely human miracle could have been greater than the one achieved in "On the Electrodynamics of Moving Bodies," Einstein's third paper—known today as the special theory of relativity. Its premise was not that everything is relative, but that some things are *not* relative: if a law of physics is truly a law, it will be true for everyone, no matter where he is or how fast he is moving. In other words, all observations and measurements will be relative to certain basic absolutes—boundaries of experience and fact that can never be seen to alter.

Maxwell's laws, thought Einstein, were really laws—they were among the most fundamental and unchanging truths of human experience. "The same laws of electrodynamics and optics will be valid for all frames of reference," he wrote. Those laws made it necessary that light should move through a vacuum at the same speed, everywhere and always.[7]

To insist on this limit was to lay down a new girder in the scaffolding of existence. Since the days of Newton and Galileo, space and time had been thought of as a rigid grid through which the bodies of the world could move. Earth might go hurtling around the sun; boats might go lurching over the sea; but behind it all there was a stable framework that defined the absolute position of each point in space. It was inconceivable that these points should ever move. They were the basis of motion itself, the still backdrop of all things.

But if Einstein was right, then that backdrop must be torn down.[8] For if a beam of light could dart across the carriage of a moving train at its characteristic immutable speed, then a watching bystander on the platform must see it moving with that same speed. This despite the fact that the train was also moving, so that its own speed should have been added to the light beam in the bystander's eyes. If light's velocity was to stay constant, time and space themselves must dilate and bend to make up the difference.[9] The result was to sweep away "absolute space" and "absolute time" along with the ether that was supposed to float through them like so much vanishing smoke. Light was not moving against some invisible background: it *was* the background. From its flow of energy, space and time emerged into view.

The Broken Boundary

To set the cosmos on this new footing, Einstein had to melt down the machinery of Newton's world. Matter itself began to lose its rigid contours, since its mass is bound up with the energy produced by its motion—and motion is relative to light.[10] The relation $E=mc^2$ arose directly from special relativity. It raised the dizzying possibility that material particles, the rigid and dependable bodies whose motion was described by Newton's physics, might simply dissolve into the sea of energy that now seemed to throb and churn beneath the surface of the solid world.

Though it seemed fantastical, this was no mere speculation. In 1920, the English astronomer Arthur Eddington hit upon a process that could generate a rush of energy from the collision of miniscule bodies, fusing together two hydrogen nuclei.[11] The total mass of the resulting helium would be less than the sum of its component parts had been before the reaction, as the excess substance exploded outward in a burst of heat and light. The sun had been pouring forth a cascade of this transmuted matter since its birth.

Every second that passed in earth's nearest star, hydrogen particles were crashing together in the millions of tons. In doing so, they passed partly through the veil that separated flesh from spirit and unburdened themselves of solid existence by way of energy radiation. The boundaries of the material world were porous; the evidence of this could be felt on the skin each dawn as sunrise flooded the world with light.

Matter could become energy. But what was "energy"? It was one of those old exiled ghosts, an immaterial spirit hovering through a supposedly material world. From this point on it becomes crucial to keep watch on the tics and habits of thought itself. The mind is embodied, and so are we: if there is anything more than matter in the world, we must know it *in* and *through* our sensory contact with matter. And so, when we try to think of abstract and immaterial things, we always *picture* them, representing their nature with some sensory image. "The mind never thinks without an image," wrote Aristotle, but images are not the same thing as the ideas they convey.[12] Einstein's own contemporary from Warsaw, Alfred Korzybski, wrote that "a map *is not* the territory it represents but, if correct, it has a *similar structure* to the territory."[13]

Arrows on a drawing of a river are not arrows in the water itself. They point in the direction of the current that flows through the water, charting a tendency that can be discerned in its movement. No one can picture the wind without envisioning its effect on something else—without seeing it in the mind's eye ruffling among leaves or sweeping through a curtain of rain, brushing across the surface of the skin. But wind is not leaves, or rain, or skin: it is a pattern of motion through air. It is not matter but something that matter *does*, a fact about how it moves. The way we think in pictures can trick us into forgetting that what we see *in* and *through* matter is not always in itself material. Objects are not the only things there are.

So when we say a word like *energy*, we may picture a pulse of glowing light or a stream of arrows from one location to another. One of Michael Faraday's most accomplished modern successors, David Tong, invites audiences to picture energy fields as a form of "fluid, which ripples and sways throughout the universe."[14] Out of necessity, we slip into talking about energy as if it were some kind of jelly or paste, soaking through the air like invisible liquid in the pores of a sponge.

But these are only pictures of energy, not the thing itself: real energy, as we encounter it in the world, is simply the capacity to do "work." Work, in turn, is force applied to move an object: it is a push or a pull that causes displacement from one place to another. Energy is power and capacity; it is potential and prediction. *This* object will do *that* under *these* circumstances: when we say something "has" energy we mean that we can foretell how it will behave—what potential lies latent within it.

Search every crevice of the object itself at any given moment, take the most minute still-frame image of its tiniest particles, and you will find no "energy" in it, no crackle of light or glow of power to correspond with what you picture in your mind. What you are picturing is not a material object to be touched or seen: it is an immaterial fact about material objects, a property whose effects can be experienced *through* the senses and their imagery, but whose essence can be known only by the mind. In the days when Aristotle puzzled over why anything moved at all, he imagined powers inherent in the nature of things themselves. Those powers are not vanished, though we call them by other names. The world is no less teeming with invisible motive forces than in the days when angels moved the stars.

Contortion

With special relativity, the sharp edges of the mechanical universe were beginning to blur. But Einstein's ambitions extended beyond even the boundaries between matter and energy. Ten years later he presented his general theory of relativity, so called because it claimed to lay down a bedrock of general laws that would hold steady for every observer under every possible condition—even under the influence of gravity.

The mysterious pull of gravity had vexed physicists since Newton. It seemed to call out across space from body to body, surging through the ether with instantaneous and irresistible power. But if there *was* no ether to begin with, then perhaps the world was not a uniform expanse filled with physical substance after all. Perhaps it was contoured by an underlying network of relationships—perhaps gravity was not a force extending across an empty plain, but a feature of the terrain itself.

General relativity recast space and time not as constants but as variables—not as given inputs into the data of experience, but as outputs resulting from the situation of the observer. What we feel as the pull of gravity is identical to what we would feel if our frame of reference were accelerating, if the whole network of coordinates in which we find ourselves were in a changing state of motion.[15] With that claim, Einstein had built the world anew, changing it from an empty grid of space containing matter into a seething ocean of fluctuating energy.

Much as Faraday suspected, the world of solid things was not built up from inert chunks of material. It was born from the forces that writhed and coiled together into ever-changing configurations. Einstein had made good on a long-held intuition that when he cleared away the painted scenery of the world, he would find not particles of matter but arrangements of energy and force: "then only field-energy would be left, and the particle would be merely a special density of field-energy."[16]

When the mind peered at last into the secret chambers behind the stage-set of the physical world, what emerged were not objects in space but interlocking energy meshes called "fields." These were unbounded expanses of infinitely small points, each one of which could be felt to exert a certain quantity of force or charge. "The only statements having regard to these points which can claim a physical existence are in reality the statements about their encounters": given an observer's position on the rolling topography of space-time, the appearance and behavior of the world around him could be predicted and understood.[17]

But the landscape *beyond* the observer's view was beginning to shift and warp as in a surrealist painting. In the unseen regions outside daily experience, particles could no longer be counted on to move serenely on the courses set for them by Newton's laws. The small gods had begun to misbehave. When space and time extended far enough, the warp in the fabric of things became apparent. Across the yawning distances that separated earth from the stars, light itself would curve along the slope of gravity, slipping out of what would have looked from earth like a straight path.

Fascinated by these possibilities, two groups of British researchers set out on an expedition to put Einstein's theory to the test. So it was that one spring day between the wars, in the eerie calm of a solar eclipse, far-flung travelers watched as starlight flooded past the darkened sun to meet the gaping eye of a camera lens. In the course of their journey, the stars' beams had wavered a few mere fractions of a millimeter out of their expected path.[18] As it shot past the massive orb of the sun, the light from distant stars had been swept into an arc as if it was rolling along the inner rim of a giant bowl. From an earthbound perspective, the gravitational pull of the sun's mass amounted to a contortion of spacetime, so that the straight lines of the light rays flexed into a curve. Newspapers published the results in November 1919, furnishing dramatic support for Einstein's theory and launching him headlong into celebrity. Just as he predicted, the boundaries of the material world were beginning to buckle and bend.

That was far from all. The whole picture of things laid forth by Newton's physics—the orderly world of bodies in motion—had begun to wash away. But the greatest upheaval was yet to come. Einstein helped set it in motion himself, though when it came fully into view he would resist it with all the considerable force of his ingenuity. The most severe blow to classical physics would not be dealt in the outer regions of space: it would strike much closer to home. All the while that Einstein was working to bring light and matter together, the atom—that supposed rock upon which all things were founded, the basic material component of all objects—was beginning to fracture into something altogether strange and terrifyingly unstable.

Breakdown

It started less than two decades after Maxwell's death. In 1897, the atom began to splinter and crumble into smaller component parts. J. J. Thomson, the brilliant director of Cambridge's Cavendish Laboratory, discovered a portion of the atom that could carry electric charge. A sealed glass tube, if emptied of air, could pass electricity from one metal plate (the "cathode") to another (the "anode"). When the anode was sealed inside the tube in the form of a cylinder, the "cathode rays" of moving electricity passed through the metal and struck the glass, making their presence known with an otherworldly glow.[19] What Thomson showed was that cathode rays were not simply beams of pure light: they had *mass.*

The electricity was carried in the body of particles, smaller than any known atom, soon to be christened "electrons." To Thomson and his colleagues this seemed like a victory for atomic theory, confirming that matter really could be pared down into discrete units with measurable mass. These "corpuscles," these tiny bodies, were really there in the translucent cathode-ray tube, gleaming against the walls of glass with triumphant light. At the Cavendish's annual dinner that year, elated research fellows sang a drinking song to celebrate: "The corpuscle won the day / And in freedom went away / And became a cathode ray!"[20]

Of course, though, an atom with parts is no "atom" at all. Far from being "indivisible," the atoms of Thomson's lab were quite obviously divided. If electrons carried negative charge, then they would naturally repel each other. To keep things from exploding outward into total confusion, the atoms making up everyday matter must be bound together with a corresponding positive charge. Thomson imagined this stabilizing power as an orb of diffuse energy, a loose field studded through with electrons like a sponge pudding dotted with candied fruits.

But at the turn of the twentieth century, under the direction of Ernest Rutherford, Hans Geiger and Ernest Marsden discovered that small bodies of positively charged mass called alpha particles could repel against the atoms in a thin sheet of gold foil, like miniature planets turning sharply in their orbits. If most of the atom was empty space, the particles should have sailed easily through the gold. Instead they would sometimes bend backward in their course, just as one positively charged body would do when confronted with another. "It was almost as incredible as if you fired a 15-inch shell at a piece of tissue paper and it came back and hit you," Rutherford recalled.[21]

Gradually an image came together, a dense cluster of positive "protons" and uncharged "neutrons" sitting as an anchor among a swarm of electrons. In the relative terms of the atom's tiny scale, the electrons swept nimbly along a distant periphery. At the compact center—the nucleus—protons and neutrons each contributed one unit of mass.[22]

The system itself hung in elegant counterpoise, and—most important—each component part was an interchangeable unit. Their behavior relied not on their color or texture, nor on any merely subjective quality. In essence, they were purely measurable packets of matter, like incarnate numbers to be tabulated and exchanged. All that mattered was their mathematical properties: how many of them there were, how much mass they possessed, and how they moved.

Vexingly, however, they did not quite move as expected. As Maxwell had once stressed, there was a hitch in the smooth course of atomic motion. According to the equations of Newton's physics, atoms should move indifferently along whatever path their collisions dictated, rolling through space like billiard balls across a table. In practice, though, there were some kinds of motion that seemed forbidden by the laws of nature.

If a gas was released into a room, for instance, its atoms could only ever spread themselves out across space and intermingle in an ever-expanding cloud. Theoretically they should also be able to reverse course and regain their original formation, falling back into a more compact and orderly arrangement. But time never flowed backward. Like the angel standing with his fiery blade at the gates of Eden, the "second law of thermodynamics" barred the journey homeward. Disarray and decomposition was the rule; order, once lost, was never recovered.[23] Why?

The best available answer was also the most disturbing one. Back in the nineteenth century, in the quiet Austrian town of Graz, the mathematician Ludwig Boltzmann had developed a theory that atoms were not *bound* to scatter into ever looser and more chaotic patterns. It was just that those patterns were enormously more likely than their alternatives, on an order of magnitude so great that there was almost no chance of observing a return to order out of chaos.[24]

It was acutely ironic that this idea should have been dreamed up by Boltzmann, the grandson of a clockmaker. For Boltzmann's statistical mechanics implied that the universe is *not* clockwork after all: the perfectly predictable world of mechanical motion was a fantasy. Atoms were highly regular, and with the discovery of their structure they were starting to seem almost perfectly numerical. But for all that they could offer in the way of power and prediction, the small gods refused to stay entirely within the bounds of necessity. The world could not be welded into a rigid course of predetermined action. Something in the system remained frenetic; some tremor was always evading neat measurement and refusing to be factored away.

The Quantum of Action

It got worse. Not only did the atoms that made up matter fail to travel in lockstep motion, but the energy that passed between them seemed to evade neat calculation as well. In Germany at the turn of the century, a theorist named Max Planck registered a disturbing hiccup in the behavior of light. Maxwell's field equations dictated that electromagnetic rays should carry energy in an unbroken stream, surging through the atmosphere until they met with a surface that could receive the energy and vibrate in response. But at the most miniscule level what Planck found was not the clean simplicity he longed for—not reality shaved to its bare essentials—but a jarring sputter in the space-time continuum.

It showed up when Planck tried to model the warm recesses of a glowing "black body," a cavity like an oven or a kiln. In a black body, heated surfaces absorb and emit energy without reflecting it, sending the energy of their temperature shooting forth in the form of light. By the classical reasoning of Planck's predecessors, the radiance that could flow out of such a body ought to have been without limit, a dark cavern throbbing with infinite amounts of light at infinitely high frequencies. In practice, the light resolved itself into a limited range of wavelengths.[25]

Instead of flowing cleanly along a spectrum of energy levels, as any wave should, light came packaged in discrete units.[26] It seemed as if its energy could only be transferred in stages, proportional to whole multiples of a certain (though apparently arbitrary) number, later known as "Planck's constant." This stubborn grain of indivisible energy could never be smoothed away. Hard as Planck tried, he could never eliminate it—it just moved from place to place in the equations like a bubble under a tablecloth that refused to lie flat.[27] Gradually Planck and his fellow theorists had to accept that the transfer of energy was not endlessly smooth: there was an indivisible "quantum," a small amount beyond which no further divisions could be made. "One will just have to live with it," Planck conceded.[28]

So just as matter's atoms were breaking down into ever-smaller portions, energy was resolving from a continuous flush into something very much like atoms. And it was here that Einstein, who saw that energy and matter could pass into one another, put forward a proposal with unsettling implications.

Perhaps the quantum was not simply a formal nuisance, a mathematical abstraction that would disappear with more sophisticated observations of the underlying reality. Perhaps the quantum *was* reality: perhaps light could be pictured not as a fluid wave but as an unbroken stream of particles or "photons," shooting through space with momentum but without mass. If these particles came with a finite amount of energy, the limitations of Planck's constant could be physically explained.[29]

But this was a monstrosity. Photons could not possibly exist: experiment and theory alike had shown again and again that light was a wave. To turn back now and make it into some kind of massless matter—a preposterous concept in and of itself—defied everything that seemed certain. But there it was, like a ghostly new picture layered on top of the old one, each competing for prominence, as the occasion demanded. Light was an energy wave—but it could also, in its smallest quantities, behave like an object. Under the right conditions, light seemed to congeal into particles of matter.[30]

Einstein's discoveries had already blended matter and energy. The next step came when a young aristocrat named Louis de Broglie suggested in 1923 that light might not be the only particle to breach the porous boundary between material and immaterial. In a note submitted to the French Academy of Sciences, de Broglie defended the wild claim that *any* particle with mass could be considered mathematically equivalent to a wave with proportional frequency.[31]

With that, the foundational categories of classical physics began to slip and melt away for good. The world's leading minds had discovered something far stranger than mere bodies in motion. The birth of quantum mechanics was so alarming that Soviet ideologues, when they heard of it, refused to accept it and condemned it as reactionary: how could History be destined to march forward toward communism if the world was not a machine?[32] But the mysteries of atomic motion remained stubbornly unconcerned with Party orthodoxy. From the depths of Greek antiquity, from the very beginnings of science, an old question could be heard echoing louder and louder, though once it had seemed settled: *just what, after all, is the world made of?*

CHAPTER 7

The Tower of Babel

It was all very well as long as we were allowed to treat the "primary qualities," as they called them—extension, solidity, figure, and motion—as included in inferred nature. But when these began to go the way of the secondary qualities; when even solidity turned into a secondary quality, it was bothersome.

—Owen Barfield, *Worlds Apart: A Dialogue of the 1960s*

A ll the greatest minds in physics could see by now that a tectonic shift was underway. They met to discuss it in the fall of 1927, during the hush of peace between the world wars. There they were in their prime, each straining like a hunting dog after a prize buck, snapping at the heels of some new breakthrough that would make sense of the quantum. There was Einstein, the electrician's son, leery of authority and ill at ease in the high society that had scoffed at him not many years before. There was Prince de Broglie, the French aristocrat of Italian extraction, youngest scion of a dynasty whose origins stretched back into the glamorous medieval past. There was Niels Bohr, the soft-spoken philosopher; Paul Ehrenfest, the boisterous New Zealand rugby lad; Erwin Schrödinger, the passionate maverick; and Hendrik Lorentz, the beloved elder statesman.[1]

They were there to bring order back into the universe. It was a matter of urgency. World War I had shown just how fearsomely consequential real scientific understanding could be—how it could give and take life, how the balance of world power could hang upon its management or mismanagement. The steel that tore across miles of train track could just as soon be sent ripping from a Mauser rifle into human flesh. Einstein's fellow German, the chemist Fritz Haber, had pulled nitrogen straight from the air to fertilize vast fields of crops and feed masses of hungry men and women. He had also concocted the chlorine gas that descended like a curse over the trenches and sent young men writhing to their graves, choking on blood and sputum.[2]

Einstein's loathing of the war effort had kept him at arm's length from his countrymen for years. But as the horrors of those days faded into memory, he and his fellow scientists could tentatively aspire to work on gentler endeavors. They came together across the primly manicured lawns of Leopold Park outside the center of Brussels, near the still water of a manmade pond. It was quiet now in the city, and the leading lights of the troubled modern world could hope for peace. But they could not agree.

Just as Europe's glimmering civilization had collapsed inward upon itself in a savage bloodletting, so now the elaborate architecture of scientific knowledge, carefully built up brick by brick since the days of Galileo, tottered and threatened to cave in. The mathematical equations of physics, which had seemed to trace in outline the very foundation stones of reality, were now delivering results that no one could satisfyingly interpret or understand. Not only could waves of energy harden into mass, but particles of mass could melt into patterns of energy. The orderly world picture of classical physics had begun to shift and bend into something unrecognizable.

At the end of the 1927 conference, Paul Ehrenfest, one of the world's greatest mathematicians, walked to the blackboard and wrote down a passage from the Book of Genesis: "And the Lord said: Go to, let us go down, and there confound their language, that they may not understand one another's speech."[3] It was supposed to be a primitive legend from the distant past: the Tower of Babel, humanity's misbegotten effort to make a name for itself. A jealous God, said the story, had descended to turn the builders' surefooted ascent into a chaos of confusion and disarray. But physics was not supposed to be vulnerable to this kind of supernatural interference. It was supposed to be a sure and rational picture of the world, as exact as it was tangibly real. *This* tower could not be shaken: its foundation was in the solid rock of the earth, and its spires reached up in graceful aspiration toward the highest heavens. Yet here were its most skilled architects, and they were as mutually unintelligible to one another as if some vengeful deity had scrambled their speech.

The Dissolving World

The problems came down to the most basic interpretations of the new physics. If a particle can be a wave, what is it a wave *of*? What medium is it moving through? In the Christmas and New Year's season of 1925–1926, Schrödinger gave mathematical form to the waves in question. Absconding dramatically to a Swiss villa with one of his several mistresses and a copy of de Broglie's PhD thesis, stuffing a single pearl in each ear to drown out the noise around him as he worked, the mathematician rummaged in the dark heart of a strange new world. The result was six papers outlining the mathematical rules governing quantum waves.[4]

Schrödinger's wave function, for which he won the Nobel Prize, is an operation that describes a set of "standing waves"—waves that form a set pattern, rather like a guitar string vibrating up and down. Schrödinger equations can produce values for the total energy of any subatomic particle, tracing its contours to describe a permanent wrinkle in the bedrock of existence. They have proven as fundamental to quantum physics as Newton's relationship between mass and acceleration is to classical systems. Schrödinger had developed an outline of the waves that doubled as the smallest known particles. But he had nothing to say just yet about what those waves were moving through.

It was the German physicist Max Born who proposed an answer: Schrödinger waves describe the probability of finding a given particle in a particular place, or possessing a particular momentum, under given conditions. Solutions to the wave function indicate how likely it is that a particle will be found at any point in space. The same goes for the particle's momentum: the equations predict which values are likely to emerge if it is measured—but none that are certain to emerge. The wave function is, in Schrödinger's words, "the means for predicting probability of measurement results."[5]

This was a monumental achievement. It was also a disaster. The entire edifice of physics—the whole Tower of Babel—rested on the faith that numbers could describe objects in themselves as they are, liberated from any error or ambiguity of human judgment. Philosophy, Galileo had written in the seventeenth century, "is written in the language of mathematics."[6] That language was supposed to be pristinely exact, parsing the joints and outlines of reality like a razor blade.

If a given star was a sphere of a million kilometers in diameter, it would remain that way even in daylight: though it may disappear from sight, its contours would stay firmly in place. To a human onlooker from earth it might glimmer or fade with the passing of night, curving in its arc across the sky. But in truth, in the regions untouched by any sense perception, the star of the heavens was no flickering pinpoint. That was only its transient effect upon the human eye, the outline it had first cast upon the confused screen of an upright ape's clouded mind. Men had risen beyond those muddled origins now to grasp the truth of things: a star was an object like any other, described by its measurable quantities, hanging firm in the space allotted to it by its geometric outlines. The thing in itself corresponded exactly and directly to the numbers that described it.

Among those numbers, the atom was supposed to be the simplest unit. It was the alphabet of the mathematical language. That is why so much rested on the particle, the small god moving rationally and eternally, enduring undisturbed beneath even the most violent change. As the component parts of the atom came into view, protons and neutrons were assigned a weight of one "atomic unit" each, as if to fix them as the basic unity that made counting things possible, the number out of which all other numbers are built.

Existence was supposed to be made by tallying up these numbered objects, each interchangeable with others of its type like
tokens or tallies, corresponding in both mathematics and experiment to fixed measurements. Here was a hydrogen atom, with one
proton and one electron. Here was helium, with two each of protons,
neutrons, and electrons. Like the stars, they were set in their courses,
positioned exactly in space—their quantities, once known, could
explain all other movements and positions. Maxwell had called
them "the foundation stones of the material universe."[7]

And yet now, at the tantalizing moment of discovery, right at
the threshold of the atomic realm, the solid particles of matter dis-
solved into mere possibility and potential. They did not, after all,
occupy just one point in space, unless a measurement of their position was taken. When no one was looking, they were not definitely
in any one place: there was only a range of more and less likely
places, a wave of options shivering through a sea of likelihoods. But
options and likelihoods are not objects; any wave that travels
through them is not made of matter. It was as if the solid ground
had fallen away with a lurch and left humanity suspended in midair.
The world beyond the edges of sight had begun to dissolve.

A Terrifying Relentlessness

In May 1925, on the spare and cliff-bound Danish island of
Heligoland, a vigorous young member of Max Born's faculty
invented a strange new mathematics that could keep pace with the
disjointed movements of subatomic particles. His name was Werner
Heisenberg, and his system was a form of matrix algebra—grids of
numbers stacked and interlocked with one another. Unlike other
atomic theorists, Heisenberg made no attempt to picture what his
numbers described: it couldn't possibly be pictured. The quantum

behavior of the atom, when it wasn't being measured, defied every category and boundary of human experience. Heisenberg's idea couldn't be visualized, couldn't be sculpted out of clay or bronze like some ancient model of the solar system or Dalton's wooden models of the atom. Heisenberg's mathematics was simply an array of relationships that predicted what *would* be seen when a human experimenter made measurements. And it worked.

The discovery left Heisenberg lightheaded, as if he was peering past the outlines of the material world into a deep underground of pure ideas: "I had the feeling that, through the surface of atomic phenomena, I was looking at a strangely beautiful interior, and felt almost giddy."[8] Sleepless, he clambered his way to a spur of rock jutting over the coast. Before his eyes, the sun rose as it always had over the cold sea. But beyond his vision—outside the range of any human sight, in the unseen world excluded from sense perception—a far more alien landscape was taking shape. There, in those new regions beyond the boundaries of what any human could see or imagine, the particles of nature burst the constraints of physical embodiment and performed unimaginable feats.

Heisenberg's mathematical latticework mapped neatly onto the heady theories of Niels Bohr, a young Dane whose hulking stature masked a gentle, soulful demeanor and an eerie habit of gazing into the clouded regions where scientific observation met philosophical speculation. Bohr worked out the steps of an energetic dance between protons and electrons, identifying the positions that electrons could occupy around the nucleus of an atom where the balance of charges would remain stable. An electron that took on energy—by being heated, for example—would flit to a new region where it could keep from surging out of control or collapsing inward to collide with the nucleus. But—and here was the inconceivable thing, the spectacular event that simply defied visualization—an electron that took on enough energy would slip from one place to another without occupying any of the locations in between. One moment it was here, the next it was there. When it fell back into its original place it would again skip nimbly over the intervening space, casting forth a burst of light in the form of Einstein's massless photons.[9]

All this frustrated the most basic intuitions of scientists and laymen alike. If anything was obviously true of physical objects, it was that they traveled from one point to another across an unbroken line of motion, touching every point along the way. They did so whether or not anyone was watching; they were not supposed to change the rules when mankind's back was turned. Yet here at the most basic level, particles seemed to jump from place to place or even hover in a range of various possible locations. The picture of the world was becoming pixelated, chunked up into tiny regions of mystery where nothing could be nailed down. Planck's constant set an impenetrable barrier around the smallest areas of space, making it impossible for the mind to picture what went on within them. In those forbidden realms, the particles that had once seemed like solid objects scrambled every natural idea of the physical world, moving and changing in ways the mind simply could not envision. Beyond a certain limit, everything went dark.

According to Bohr, this was entirely to be expected. His explanation was devastating in its simplicity: what can never be seen, can never be visualized. The unobserved behavior of particles can, of course, never be described in terms of human observation. Concepts like "location" and "motion" are not in the end purely numerical facts, independent of human experience. They are descriptions of how humans can perceive and measure the outside world. Particles exist outside of us, but *all* the terms we use to describe them—even the mathematical ones—refer to our experience of them.

Quietly but doggedly, Bohr insisted on stripping down particle physics to a description of how the outside world was likely to affect human observers. "However far the phenomena transcend the scope of classical physical explanation, the account of all evidence must be expressed in classical terms," he argued. Heisenberg's mathematics, Schrödinger's probability waves, Einstein's photons, and Bohr's own electron orbits were all impossible to picture, because what they described could never be experienced.[10] The outside world was real, and facts about it could be known. But they would always be known through observation and measurement, filtered through the shapes and forms of a conscious mind.

Quantum mathematics gave information about how a man *could* experience a single particle once he measured it. But as for what was happening to the particle when no man was looking—it could not even be spoken of. Particles were not "in many places at once" when described by the Schrödinger wave; they were not in *any* particular place, because no one was experiencing their location. What's more, Heisenberg's equations showed that experiencing their location put an inherent limit on knowledge of their momentum, a so-called "uncertainty principle" that set bounds to which physical properties could be resolved from potential to reality at any given time.[11]

The idea appalled Einstein and unnerved Schrödinger, both of whom came up with one thought experiment after another to try and prove that something outside the realm of human experience could be known. When they met at conferences, they haggled over these puzzles for hours, late into the night, searching for a way around Bohr's obstacles—but to no avail. Heisenberg remembered the unyielding intensity with which Bohr defended the boundaries

of human knowledge: "for though Bohr was an unusually consider-
ate and obliging person, he was able in such a discussion, which
concerned epistemological problems that he considered to be of vital
importance, to insist fanatically and with almost terrifying relent-
lessness on complete clarity in all arguments. He would not give up,
even after hours of struggling."[12]

Destroyer of Worlds

Bohr and his collaborators were used to pressing beyond the farthest frontiers of what was known about the world. But now Bohr
was forcing them up against the limits of what could *conceivably*
be known, the hard barrier of the mind itself.[13] Beyond the border-
land of human experience, where no measurement could be taken
and no observations made, the normal framework of space and
time simply ceased to apply. When subatomic particles did slip
back into view—when they made an impact on the world of human
perception through a measuring device—they bore clear marks of
having behaved in ways that could not and would not make sense
to picture.

The mind is like a sheet of fabric with images of space and time
sewn onto it in needlepoint: when they pierce into the domain of
human experience, atoms and particles resolve themselves into neat
objects traveling in continuous lines. But underneath the canvas,
the picture collapses into a tangle of shifting threads. Out in the
wilds of the unseen realm, isolated particles cease to be particles
altogether in any meaningful sense of the term; they lose all definite
sense of place and billow outward into a cloud of possible
locations.

Send a single electron or photon traveling through a narrow slit
in a barrier, then through two more slits on the other side: it will

not pass neatly through in a straight-line trajectory. Instead it will scatter into one of several regions predicted by Schrödinger's wave function, as if its many possible trajectories had gone on lapping against each other until they crashed and broke against the rock of the visible world.[14]

Or shoot a photon at a device called a "silvered mirror," which has a fifty-fifty chance either of letting the light pass through or of reflecting it onto a different path. Then set up regular mirrors in the way of both possible trajectories, bringing the potential paths back together at a final silvered mirror. The one photon will then behave as if it has taken *both* roads at once, "colliding with itself" at the last mirror and ending back up on a single course.

Or, more confoundingly, subject an electron to the influence of a magnet and it will exhibit behavior that can only suggest it is making two rotations in one, spinning twice rather than once before it ends up for the first time back at its original angle—a meaningless statement, impossible to visualize.[15] The words themselves for describing motion and shape break down; what cannot be perceived cannot be stated in terms of human perception. Quantum equations don't describe the outlines of a world we can see and touch: they describe the limits where things cease to become tangible or visible. The world beyond those limits is not made of solid objects.

Yet it is real. Any teacher in your average American high school will point out areas on the board using a LASER (Light Amplification by Stimulated Emission of Radiation), a device which creates a cascade of photons pouring forth in a narrow beam from the precise transition of electrons between energy states. These quantized glimmers can be honed to pinpoint accuracy and shot in coded patterns along the fibers of a modern internet cable, a flicker of information whipping across space in the form of light. Computers may soon wildly outstrip their current speeds using quantum bits ("qubits"), which can register not only ones and zeros but both also a simultaneous blend of the two, occupying a "superposition" of two binary states at once.

And of course, the world's first quantum physicists received devastating confirmation of their own theories when, to the mixed horror and amazement of a watching world, two clouds of wreckage hovered over Japan and brought an end to World War II. Firing neutrons at the core of uranium and plutonium atoms, the scientists of the Manhattan Project built the world's first atomic bombs by sending the bolts and rivets of the universe skittering apart. The grains of the world came tumbling through the crevices of space, as atoms at last gave up the bonds that made them whole in a thunderous blast of energy and became death, destroyer of worlds.

The Reappearance of Things

It was not just cities that the quantum revolution wiped away into nothing. An entire picture of the world was being erased: the picture of the mechanical universe, with its solid moving parts churning on unseen, was beginning to fade and dissolve. For some of the twentieth century's greatest thinkers, this prospect was too much to bear. Einstein refused to countenance what he called the "ghost waves" and "dice games" of quantum mechanics: somewhere underlying all of it there must be a solid mathematical picture, "a theory which describes exhaustively physical reality, including four-dimensional space," whether observed by a human onlooker or not.[16]

Yet it was becoming clear to everyone, as Schrödinger put it in 1931, that "the mathematical apparatus derived by Newton is inadequately adapted to nature."[17] At Bohr's relentless prompting, the pioneers of quantum physics were forced to grapple with the possibility that their discoveries might—in the words of Schrödinger's translator and interviewer James Murphy—"reduce the last building stones of the universe to something like a spiritual throb that comes as near as possible to our concept of pure thought."[18] Bodies in motion were not the heart of things: they arose out of a darker and more mysterious well, resolving before human eyes into the familiar shapes described by geometry. Perhaps it was *only* when they could be seen by a conscious observer that "objects" made any sense at all.

This was an ancient truth. It had been hinted at in scripture and wisdom literature all along. But it also represented the end of a centuries-long quest. The scientific revolution had begun in the hope of going beyond appearances, puncturing the fantasies of the human mind and seeing things as they "really are." Galileo and Descartes had aspired to gaze past all the merely subjective qualities of the material world, peeling away color and texture and sound until only the raw mathematical facts of position, size, and location were left. But now it is becoming clear that even those supposedly pristine numerical attributes are partly indebted to human perception. They are not arbitrary: any two healthy observers can match up their experiences and agree on the location of a tree or a house. But even concepts like "location" can only accurately be used to describe the smallest objects once someone is there to do the describing.[19]

It could never have been otherwise. From the beginning, mathematics was only ever designed to "save the appearances." It was an abstraction derived *from* and *about* human experience, created to *explain* and *apply to* human experience. The most optimistic impresarios of the scientific revolution made more daring promises on behalf of numbers, claiming that they could trap the universe in amber and freeze the world into a perfect interlocking mechanism of material objects. But the world is not made only of material objects. It is made of the meeting between mind and matter. It is this human encounter with the outside world that brings it shimmering into an array of form, color, and light.

The human mind—that supposedly primitive and dispensable screen of illusions—is far more fundamental than was once assumed.[20] It is not an obstacle, throwing up its dreamlike fancies of selfhood and spirit into the way of hard numerical truth. It is a vehicle, bringing the potential of the unformed material world into definite reality. Things are not divided into "primary" and "secondary" qualities, split between the hard facts of matter and the vaporous deceit of human experience. Human experience is *part of* the world, which is built up from the raw potential of existence into the panoply of color and light that passes before our eyes. The reign of the small gods is over, and their Tower of Babel is fallen. We are left at the center of the universe, helping to grant it life.

PART 3

All Things That Were Made

Light in the world—
World in the mind—
Mind in the heart—
Heart in the night.

Pain in the day—
Strength in the pain—
Light in the strength—
World in the light.
—Owen Barfield, "Meditation"

CHAPTER 8

Light of the Mind, Light of the World

The Atoms of Democritus
And Newton's Particles of Light
Are sands upon the Red Sea shore,
Where Israel's tents do shine so bright.
—William Blake, "Mock On, Mock On, Voltaire,
Rousseau" ll.9–12

The world is made in the meeting of mind with matter. This lost truth is coming back to us now. It was always hovering in the background like a forgotten melody, underscoring the great adventure of human knowledge. To investigate the world, you have to see the world. And so, the world we investigate is the world as we see it. But we grew so absorbed in our investigations that we forgot the melody was there.

Gradually, over centuries, we lost sight of ourselves. We became so focused on bending the world to our will that it came to seem like a dead thing in our eyes, an object outside of us to be toyed with. And then, like the worshippers of stone idols, we came to see ourselves as objects too. "Eyes have they, but see not," says Psalm 115, referring to the austere statues that loomed silently above their cowering subjects in the temples of the false gods Baal and Marduk: "they who make them will become like them." Those statues are fallen, but today we come to new temples looking for the same old glories: power over nature and freedom from death.

That is what we hoped for when we turned the world into a corpse. Francis Bacon predicted that once nature had been wrestled into submission and subjected to questioning, her interrogators would become "Stewards of Divine Omnipotency and Clemency, in prolonging and renewing the life of man."[1] Total control over human life: the inheritors of Bacon's dream have never seemed closer to this goal. "What you are talking about now is like a second industrial revolution," says techno-futurist Yuval Harari, "but the product this time will not be textiles or machines or vehicles or even weapons. The product this time will be humans themselves."[2] If nature works like a machine to make man, then the man who builds the best machine can make a new nature, and a new man.

But what kind of man? If the world is an object, then so are we, and here the grim prophecy of the psalmist comes true. Those who treat the universe like a heap of material will come to treat human beings as hunks of flesh to be rearranged at will, medicated or cut to pieces by the strong. If this process reaches its end then humanity will literally disappear from view, absorbed into an unfeeling ocean of churning atoms. And since matter without a witness is matter without form, the end of human consciousness will be the end of the world. If we are objects, then we are nothing.

Yet we are *not* objects, or else we could not see. That is what quantum physics has reminded us of. In the fourth century AD, when Saint Augustine went looking for his creator, he peered at the stones and the sea until they seemed to answer back at him: "we are not your God. Look higher."[3] Now atoms themselves are crying out like the very stones, revealing by their behavior that they are not enough to make the world. It is meaningless to even speak of matter unless we talk about the effect it has on a human mind. To understand the world, we have to realize that we are the ones who see it.

And that's why it was no accident that quantum physics began with an age-old puzzle: what is light? The very nature of the question forced us to consider ourselves. The only reason we can ask it at all is because light is something we *see*. That's how we know to ask about anything: because we experience it somehow. We ask why stones fall and planets fly because we see them do it; we wonder what blood is because we feel the life draining from us when we bleed.

But light has a special kind of property. It is both a visible thing, and also the cause of vision. The monks that scraped out copy after copy of ancient manuscripts had to keep their candles burning to reveal the words. It was thanks to reflected sunlight that Galileo could see shadows in the craters of the moon. And the voyagers who stood beneath a darkened sky to watch matter bend space—they too could see that Einstein had been right only because of the glimmer that arced across the heavens from the distant stars. Light isn't just one of many things we see. It's also the reason why we see anything.

This simple fact has given new meaning to a very old idea: there are two kinds of light. There is the kind that we see, and the kind that makes us able to see. The light of the world, and the light of the mind.

Twin Flames

The light of the world streams from outside of us and into our waiting eyes; it is beyond and distinct from us. But the light of the mind comes from within, rushing out from the eyes to meet its counterpart in the surrounding environment. For many centuries it was a common assumption that both kinds of light were formed of the same stuff. They were twin fires, burning inwardly and outwardly. In Plato's *Timaeus*, the title character unfolds a tale from the days when mankind was made: "The gods siphoned off a portion of fire that cannot burn but only gives forth a gentle light, akin to the light of each day. And for it they created a brother light—the smooth, pure, and delicate flame that flows forth from our eyes." When the eye's flame mingles with its companion in the outer air, Timaeus hints, it forms a channel between the objects of the world and the mind of the watcher. "The result is the mode of perception we call 'seeing.'"[4]

Gradually it became clear that the eyes could not really be shining with their own fire—at least, not fire of the same kind as burned in the outer world. Ibn al-Haytham, an ingenious mathematician who served under the Islamic lords of eleventh-century Egypt, proved this to himself when, according to one account, he over-promised on an engineering contract and provoked the stern displeasure of the deranged caliph Al-Hakim bi Amr Allah. Rather than answer for the full consequences of his failure, the story goes that al-Haytham put on a show of madness and got away with house arrest.[5] By day he labored over a new theory of vision—but when night fell and his chambers grew dark, he found no ambient glow shining forth from his face to illuminate the surrounding walls.

In his *Book of Optics*, al-Haytham picked apart the "extromission" theory that fire comes out of the body, replacing it with pure "intromission": whatever flame or beam came from the world to lighten the mind, it all flowed from the outside in.[6] The components of this idea were already latent among the theories of Aristotle, Euclid, and Ptolemy—gathering them together, al-Haytham waved away the sham flame that had once seemed to burn behind every pair of human eyes.

But still, it made no sense to imagine that the eyes did *nothing*. If light poured into the mind from outside, it did not simply pool there like liquid gold. The light that came in from the world was altered in the mind, molded from raw color and brightness into the shape of visible objects. Aristotle argued that to see something was to receive its form into the soul without its matter; the bird that wheels across our field of vision against the sky's backdrop isn't simply one more splash of color. She is a distinct creature, and the light that reaches us from her is shaped by the mind into a form we recognize.[7] Al-Haytham thought that colored light was parsed and shaped by the "faculty of judgment" (*al-quwwa al-mumayyiza*), which makes sense of objects and separates them out into coherent wholes.[8] Light in the world was only half-formed, a message waiting to be decoded by the human mind that could transform it into sight.

The light of the mind was not a physical fire for al-Haytham, but neither had it gone dim. For reality to be made manifest, the mind and the world must meet.[9]

Nightfall

But was it reality, after all? Whatever can be seen by human eyes can be doubted: what appears today may disappear tomorrow, or look different from another vantage point. "Mortals know nothing," says a wry goddess to the seeker of truth in a poem by Socrates's older contemporary, Parmenides: "they wander helpless, with divided minds, like blind and crippled men . . . in their judgment one thing can be the same and not the same."[10] There has always been the sickening possibility that the experience of the senses is mere deceit. "Whoever tries to learn about the objects of the senses never learns anything," says Plato's Socrates. "There is no knowledge of those kinds of things."[11]

So perhaps the light of the mind is just empty shadow play. Perhaps what we see is mere confusion, the light of the world distorted by its journey down the tangled pathways of the senses. As the Middle Ages reached their end and modernity dawned, these doubts began to multiply like demon spawn. Pages of conflicting opinion poured forth from the printing press; questions and accusations from Protestant reformers chipped away at the unity of the church; war and faction split Christendom into jagged pieces. Truths that once seemed absolute no longer commanded anything like general agreement. Even the earth could not be relied upon to stand still at the center of the universe.

The modern scientific revolution was born in the midst of this confusion, as a desperate effort to secure some solid ground of truth again. If humanity is doomed to be torn apart in dispute and error, reasoned philosophers in these troubled days, then knowing reality must mean clearing away every merely human aspect of knowledge. Pythagoras had taught that numbers alone were the same everywhere and always: maybe, then, the bare facts and figures of calculation could be trusted to contain unadulterated truth. Maybe it was possible to seize upon the light of the world, unfiltered by the mind.

This was the hope behind Galileo's eager assertion that philosophy is "written in the language of mathematics." It was the impulse that led Francis Bacon to proclaim that the real causes of all events are "infinitely material in nature." If they are material then they are measurable, and so capable of being fixed with absolute certainty.[12] With the triumphant arrival of Newton's *Principia Mathematica*, there opened an enticing route to knowledge that was apparently pure and simply mathematical. And with that, the light of the mind began to grow dark.

For then it seemed the truth of things could be known without the interference of any merely human delusion. Reality, the new prophets announced, was not shaded. Nor was it textured, or clouded over with any of the phantom qualities that man alone could see. Those qualities were fleeting products of a deeper and more permanent truth, a world of numbers and of certainty. Even light itself, with its fabulous array of colors, could be understood this way: after the *Principia*, Newton's greatest triumph was his *Opticks*, in which he showed how white light from the sun could be refracted through a prism and forced to spill out its secret colors for an experimenter to see.

With the rainbow pinned and splayed across a wall, Newton made it possible to argue that light's many hues were not really qualities at all but quantities: each color corresponded to a different kind of ray, whose angle of refraction through the glass of the prism was slightly different than that of its fellows. The shades of light "do differ in degree of Refrangibility, and that in some certain and constant Proportion," wrote Newton: it was a matter not of aesthetics, but geometry.[13] For true explanation and absolute knowledge, the accidents of human perception were of no use—in fact, they were a positive hindrance. Truth depended on undoing the influence of the mind.

As the centuries wore on, being human only came to look more like a handicap, a rude imposition on the stately domain of facts and figures. "Philosophy will clip an angel's wings. . . / Empty the haunted air. . . / Unweave a rainbow," groaned the poet John Keats. Slowly but surely, it seemed that the atmosphere around mankind was being evacuated of its life and color.[14] Eventually the followers of Darwin thought they could show that the mind itself was formed by chance and the laws of brute survival. So, what could an overgrown monkey expect to add to the pristine world of truth? Light, as Maxwell proved, consists in measurable energy fluctuations. There is plenty of it that no human being can see at all. It became conventional wisdom that what we do see, we see as interlopers—uninvited guests at a royal pageant whose movements we do not fully understand. It is now supposed that our perceptions come to us as if by accident, contorted and confused by the structure of the brain. There is no light in the mind, no fire behind the eyes: the mind, we're told, is an *obstacle* to light, which in truth is a purely material body—like everything else.

What We Cannot See

The consistency of rigid cause and effect, the simplicity of a world boiled down to its elements: all this, plus untold power besides, seemed to lie within reach of whoever could clear away the fog of mere humanity and see the universe in terms of bare mathematics. The fruits of this approach have been astonishing. "We owe to it, up to now, our independence, much of our security, our psychological integrity and perhaps our very existence as individuals," wrote the philosopher Owen Barfield of the attempt to withdraw from the world and study it as something separate from our human nature.[15] But this retreat has come at a price. It has left us no route *into* the world. We stand outside of it now, dividing up the pieces of the things we see, splitting existence into smaller particles without ever reaching below the surface.

What we have gained is the key to divination that astronomers and prophets have sought since Babylon, the pattern of events that makes us able to predict and control them. We are like "a clever boy," writes Barfield, "who knows nothing about the principle of internal combustion or the inside of an engine" but who has nevertheless learned "to move the various knobs, switches, and levers about." We know how to drive the car, and for this reason we flatter ourselves that we know what the car is. But that's a fallacy: we have gained "dashboard-knowledge" and confused it for "engine-knowledge."[16]

That is why the mechanical way of thinking has exhausted itself. We have ground up against the inescapable fact that we are *here*, and the observations we make are all, by definition, human observations. However regular, however mathematical, they are still the product of a living human mind. Facts and figures, real as they are, don't live outside of the soul that computes them. And *this* is what quantum physics forces us to confront.

For when we had chased the light of the world down to its tiniest motions, when it seemed as if Maxwell's equations were on the verge of dissecting radiation and sorting it neatly into another set of rules, then the light of the mind flickered on again. As the smallest tremors of energy wavered in between a state of solidity and flux, the possibility dawned on us that *we*, the watchers, made a difference in the things we watched. Einstein's photons could knock electrons from their settled regions around a nucleus; electrons could rock like waves across an array of possible positions; and it was the experiment, the observation, that would resolve these many possibilities into one reality.

The bare facts of physical bodies that once seemed so independent of us—their motion through space and their position at a given time—these realities do not, in the end, live altogether outside the mind. It's *our* eyes that see them and our minds that record them, noting their positions on a chart or following their trajectory across the heavens. "An observation describes not an 'I-It' but an 'I-Thou' relationship," writes the neuroscientist Iain McGilchrist. "We are not distinct from—*over against*—Nature: we emerge out of, live within, and return to Nature. In a sense we are Nature herself reflecting on Nature."[17]

The facts of the physical world are far from arbitrary—we can't simply wish them to be other than they are. But they are what they are, and how they are, because of what *we* are: even physical facts are facts about how things look *to us*.

The light of the world meets the light of the mind. The world we see is gathered together and given its shape by human consciousness; it can't possibly be otherwise. A world stripped bare of human experience is no world at all. It is a painted stage set, a mute idol giving off the appearance of life. We do not stand outside the world to see it "as it really is"—we enter the world so that it can *become* what it really is. The light of the mind is not a material fire, as Plato's Timaeus pictured it to be. But the light of the world is not "simply" material either: reality as we experience it is a meeting of mind with matter.

This was a secret hidden from the foundation of our modern world. There had been intimations of it all along, hints and warnings that the light of the mind could never be snuffed out altogether. A little more than a hundred years after Newton published his *Opticks*, a rival theory of color appeared. Its author was Johann Wolfgang von Goethe, a man adored in German high society for the heart-rending passion of his novels and plays. Not satisfied with the charms of celebrity, Goethe spent his mature career yearning after glimpses of the meaning that throbbed at the living core of all things: "Whatever passes away / is only an allegory," he wrote in his masterwork, *Faust*: "The incomplete / Becomes reality here."[18]

Using just about every conceivable medium, Goethe pleaded with his sizable audience to feel the lifeblood pulsing through the veins of the world. Life to him was a parable, a drama, a painting: it could not be split apart and subdivided except as an abstraction. "I can certainly put together the individual parts of a machine made of separate pieces," he observed, "but not when I have in my mind the individual parts of an organic whole, which produce themselves with life, and are pervaded by a common soul."[19] To carve existence into isolated portions may be useful for the sake of analysis. But the resulting splinters are less real, not more, than the original whole: they may become cleaner, but they also become more sterile.

Dry wood makes no sense except as the limb of what was once a tree. The atom makes no sense except as a point in the network of relationships that presents the world to us. The fullness of truth is not best revealed under lab conditions.

This was Goethe's complaint when it came to color and light: "we should think of both as belonging to nature as a whole, for it is nature as a whole which manifests itself by their means in a special manner to the sense of sight."[20] The rainbow is not first and foremost a diffraction of geometrical rays. It is first and foremost a rainbow, an array of colors spread forth before the eye of a watcher. The trajectory of electromagnetic waves, their impact on the rods and cones of the eye, the way their path is diverted by the glass of a prism: all this is the premise and the condition of the thing itself, like sheet music ready to become a song when the orchestra plays. A mathematician can say when a rainbow will be visible, and where an onlooker should stand when the sun's light wavers through the damp air of an ending storm. But only the onlooker, in whose mind the colors themselves find their existence, can say what a rainbow *is*.

Singularity

In the quantum age, even the position of particles in space has revealed itself as more than a merely physical property: it is a relationship, a way of describing how the outside world affects us. This too was foretold. "No shape or mode of extension that we can have any idea of—in perceiving or imagining—can be really inherent in matter," wrote Bishop Berkeley of Cloyne in the eighteenth century, pressing against the embarrassing weaknesses in the new science.[21] "It is from the human point of view only that we can speak of space, extended objects, etc.," wrote Goethe's older contemporary Immanuel Kant. "Space is nothing else than the form of all phenomena of the external sense." The *phenomena*, the appearances of things to us, are what we order and explain with our mathematics. When things do *not* appear to us, we cannot describe them in their barest particulars except as potential: a map of how things *can* appear, much like the one described by Schrödinger's wave functions.

To rescue our science and our philosophy alike, to carry both forward, we will need to reconnect the light of the mind with the light of the world. That means drawing from old wellsprings of knowledge, restoring an integrity that was lost when we cut ourselves off from the source of things. There are ancient voices calling to us now, forgotten echoes of a truth whose urgency grows more vivid by the day: "You are the light of the world" (Matthew 5:14). Humanity, placed in the role of watcher and guardian of existence, gives final shape and form to reality. Whether the earth revolves around the sun or the sun around the earth, where *we are* is the center of the universe. This is an old teaching, forever new. It is giving forth fresh meaning in the quantum age.

"You are the light of the world." And again: "The eye is the light of the body. If your eye is whole, your entire body shall be full of light. But if your eye is deformed, your whole body will go dark. And if the light within you is darkened—how great is that darkness!" (Matthew 6:22–23). Twenty centuries ago, the word was spoken: if the human view of things is lost, the world will go dim. The things around us will become mere objects, and humanity itself will be just so much material to be shaped and molded at will. Genes rewired, bodies hacked to pieces, souls drugged into a stupor: how great is that darkness.

There is another way, though it lies along a forbidden path. After the seventeenth- and eighteenth-century age of "Enlightenment," it became increasingly common to associate religion with obscurity and ignorance. To this day, lodged deep in the psyche of any self-respecting rationalist, there is a residual mistrust of anything remotely supernatural. There is an impulse to recoil at the sound of footsteps coming from a church, as if cowled initiates may emerge at any moment to shackle science with arcane prohibitions. Even the advocates of the spiritual outlook present it with some embarrassment, looking for ways to give it as friendly and unthreatening a gloss as possible. Even the shepherds are sheepish.

Though this attitude has become increasingly reflexive, it has also become increasingly baseless. After all, it's not the priests but the technologists who talk like clerics now, foretelling a great transformation of mankind in a blaze of avenging fire. A moment of singularity is coming, they say, when the logic of machinery will outstrip any human attempt to control it—when the world will shed its skin and emerge reborn. "The greatest catapult for humanity into a new species lies just beyond the event horizon of transgenderism," writes the transgender technologist and magnate Martine Rothblatt in *From Transgender to Transhuman*. "First comes the realization that we are not limited by our gross sexual anatomy. Then comes the awakening that we are not limited by our anatomy at all."[22]

Beyond this threshold lies a universe broken free from all its present forms and boundaries, as the flimsy scaffolding of the human mind collapses and data flows unhindered through limitless circuitry. "The implications include . . . immortal software-based humans, and ultra-high levels of intelligence that expand outward in the universe at the speed of light," predicted the computer scientist Ray Kurzweil: in a blinding wash of cold white brilliance, the contours of the universe as we presently see them will disappear.[23]

"You cannot even begin to imagine what the consequences will be," says Yuval Harari, "because your imagination at present is organic. So if there is a point of Singularity, by definition, we have no way of even starting to imagine what's happening beyond that."[24] We cannot imagine it because "we" will not be there: whatever minds or shreds of mind flash through the wastelands of the synthetic world, they will not experience things as we do—if they experience anything at all. To prophesy a post-human world is to prophesy Armageddon. And when the end of the world is on the table, it's no longer a question of faith versus reason (if it ever was). It's no longer a question of science versus religion, worship versus intellect. It's a question of *which* forms of religion will direct our use of science.

One religion tells the story of the singularity, a story that is reaching its conclusion: humanity came suddenly and randomly from the formless slime to build wonders, and soon our use will be spent. Soon at last we will restore the ruined parapets of Babel and find ourselves absorbed into a universal building project, a world machine.

But there is another story, an older one, no less magisterial than that of the singularity and no farther-fetched. This older story does not end with the evaporation of our souls like mist—instead it begins with a world prepared to find fulfillment at the arrival of the thing called man. It is now a question of which stories, and which prophecies, are true.

A Second Sun

The old story goes that there was never a world without mind. One mind made both heaven and earth, but "the earth was formless and void" (Genesis 1:1–2). In that darkness, flesh and spirit brooded together: "the spirit of God hovered over the face of the waters." There was then a private communion between matter and the one mind, unreachable by our understanding, indeterminate as the motion of a particle before we measure it. That first creation is shrouded from us by our very nature, refusing to answer to the categories with which our minds organize experience. It was and is an intimacy that we are not invited to penetrate—the universe in a state of undress. Still today, when we ask what photons look like or where electrons are in their orbitals, we come up against that locked door behind which God and nature speak in hushed voices, alone. The earth must put on form before it can be fit for us to view.

And to endow the earth with that form, the one mind spoke: "Let there be light" (1:3). Then there was light, and things began to take on shape: time began, dividing into days (1:5). Points of light became visible in the sky, "for symbols and for markers of the seasons" (*l'otot u l'mo'edim*, 1:14–15). And everything that was made was seen: the final seal of existence that God laid upon the earth, and the plants, and the heavens, was to look at them and call them good.

In a purely physical sense, it may be literally true that light holds the world together. Even in the farthest reaches of space there is a "cosmic microwave background," a form of radiation that is thought to have endured from when the universe began. These distant glimmers of original creation can get "entangled" with any other particle they meet. This means the knowable information about the light gets bound up with the knowable information about the particles it interacts with, limiting those particles' possible trajectories and ultimately, should they ever reach a human measuring device, helping to define their position in our sight. This cascade of interaction, this network of relationship built up from the smallest tremors of ancient light, may be the reason why a world which begins in indeterminacy can appear to us as the solid universe we move through.[25]

But there is a far deeper meaning to all this. Material reality is only an outward sign of a more profound truth. It has been said since antiquity that there are really not one but two suns: the invisible one, which makes reason and knowledge possible, and its physical copy, the visible sun we see shining above us.[26] "God the Father generates love by means of His heart, and the sun symbolizes His heart. It is in the external world a symbol of the eternal heart of God, which gives to all beings and existences their power."[27] So wrote the unlikely mystic Jakob Böhme, a cobbler who saw visions at the dawn of the seventeenth century and the brink of the scientific revolution. Since we can never experience nonphysical things directly through the senses, physical things must serve as "signs and markers," as imagery of spiritual things. And the experiences we have of physical light—the fact that it gives form to the universe, that it meets with our eyes and undergoes a transformation, that it reveals to us the outlines of things waiting to be made manifest—all these facts are clues about the nature of the light that made the world.

For the universe itself is created when consciousness gives shape to time and space. The world was formed when a formless earth received illumination from a mind that saw it and named it: the claim of Hebrew scripture is that the first light did not come out of matter into mind, but out of mind into matter. That is the contrary vision that now rises to rival the vision of the machine world, in which every mind is the accidental product of matter and the passive recipient of a merely physical light. Here is the answer to the cold dread of an inhuman future: from the very wellsprings of existence, it is impossible that the world could have existed without the warm light of a living mind.

And it was in this world, as its completion and its consummation, that man was made in the image of God (Genesis 1:26). The universe is not entirely complete unless the human mind experiences it: this is now the teaching of science and scripture alike. It is no less realist for being spiritual, and no less majestic for being proclaimed by the humblest particles of creation. "In the soul of man there are innumerable infinities," wrote the seventeenth-century clergyman Thomas Traherne. "The ocean is but the drop of the bucket to it, the heavens but a centre."[28] Stars and planets, ice and fire, water and earth: it is all ultimately brought to fulfillment *within us.* Whatever technological achievements may come, whatever distant reaches of space we may explore, this will remain our privilege and our sacred duty: for the universe to be what it is, we must be there to see it.

CHAPTER 9

Creation

He showed me something else, a tiny thing, no bigger than a hazelnut, lying in the palm of the hand, and as round as a ball. I looked at it, puzzled, and thought, "What is it?" The answer came: "It is everything that is made." I wondered how it could survive.

—Julian of Norwich, *Revelations of Divine Love*,

First Shewing, Chapter 5

T he light of the mind has been dimming for centuries. Even as we begin to see our own fingerprints on nature, even as the oldest voices come murmuring through the newest discoveries of science, we are not yet ready to accept the knowledge that we are cocreators of the world. By now, the rediscovery of our own divine commission comes as an unwelcome threat to everything we've been taught to take as settled. The classical picture of the world that emerged from the scientific revolution—a picture of bodies in orderly motion, safely independent of humanity—was once an exhilarating new tool for adventure and discovery. It has now hardened into a brittle dogma that prevents us from seeing what we most desperately need to see.

To really internalize what we are learning about our place in the order of things, we will have to contradict our own deep-seated instincts. We've been trained to take a mechanical image of nature as the truth about nature itself. The universe, we reflexively assume, is an unfeeling contraption built by no one. Any departure from this imagined truth feels like a return to the days of witchcraft and unreason. Putting humanity back at the center of things amounts to nothing less than insubordination or even heresy. The forces of prejudice and suspicion arrayed against a supernatural revival are as strong today as any church doctrine. A new establishment holds sway now.

In the canonical mythology put forward by this establishment, no story is more sacrosanct than the story of how the universe emerged out of nothing. From a pinprick of matter and a burst of energy, the world was born: that has been the prevailing legend ever since the interpreters of Einstein used his blueprints to set time running backward. Already before Einstein, in the eighteenth century, Immanuel Kant and Pierre-Simon Laplace thought they could see the trajectory of the solar system sweeping back to its origin point in a spinning cloud of incandescent gas—a nebula whose momentum had drawn it inward until it was clothed in the shape and substance of the heavenly bodies.[1] But Einstein's theory had the potential to extend much further. Guided by the postulates of relativity, the mind's eye could peer into the past until it reached the shrouded antechamber of time where even high priests feared to tread. At long last, it seemed, the doors could be thrown open on the birthing room of the universe.

Fitting, then, that a cleric should do the honors. Georges Lemaître was as pious a Catholic as any—a priest as well as a physicist—a man for whom scientific research sat comfortably alongside biblical faith. As far as he was concerned, the parallel tracks of these two paths to truth need never meet: "Relativity is irrelevant for salvation," he told one interviewer. "I have too much respect for God to reduce him to a scientific hypothesis." It followed that the logic of science could unfold entirely of its own accord.[2] And gradually, over the course of the 1920s and 1930s, that logic led Lemaître to an origin point for the universe governed entirely by physical laws.

Rifling through archival telescope images at the Harvard observatory, Lemaître watched as distant galaxies moved farther from Earth year after year. Spacetime itself must be expanding, he thought, its contours elongating outward.[3] That meant he could use Einstein's equations to describe this process in reverse, inferring what had happened when the world was still new. And with that, the man of God set his mind sailing through eon after eon of the galactic past.

Lemaître's mathematics undid billions of years' worth of cosmic growth in a matter of instants, until the farthest reaches of the universe came rushing back from their distant borders to converge upon a single point. It was called a singularity, though quite unlike the one that now hovers as a prospect in the post-human future. This singularity sat far in the past, the lone gathering place of everything that had ever been or would be. Here was the world in embryo, a dense "primeval atom" from which the cosmos had sprung at a moment before which nothing could be. "If we imagine going back in time," wrote Lemaître, "we approach this singular instant, the instant that has no yesterday because yesterday, there was no space."[4] The echoing stillness of this silent beginning was exploded by what is now called a "big bang," a cataclysmic blast of the unseen force that would come to be known as "dark energy."[5] From its birthplace in the time without time, matter was propelled outward into the great dance called the cosmos.

Masterful as this argument was, it brought with it a certain dark foreboding. The problem was that strange primordial instant, that day without yesterday when all things were gathered into one. If Lemaître had brought humanity to the verge of all material existence, then he had also confronted mankind with the precipice beyond which no merely human thought could go. Any further and reason itself would lose its footing, hurtling off the edge of time into a chasm of eternity where mortal rules need not apply. Critics like the astrophysicist Fred Hoyle, who coined the term "Big Bang," gave voice to the looming question: if the universe had a beginning, what force or power lurked beyond its starting point? Had the churchman Lemaître cracked open a back door for some all-powerful deity to slip in again, demanding recognition from the scientists who had done so much to describe the world without his help?

Bounded in a Nutshell

Years later, Lemaître recalled Einstein's misgivings about the big bang hypothesis: "from the point of view of physics, it seemed to him absolutely abominable."[6] Shakespeare's Hamlet moaned that "I could be bounded in a nutshell and count myself a king of infinite space, were it not that I have bad dreams" (*Hamlet* II.ii.273–75). Here was the universe bounded in a nutshell, its starting point wrapped neatly within a microscopic span. And yet the framers of this satisfying picture found themselves haunted by bad dreams of God.

Lemaître pleaded innocent to the charge of underhand theology. "The hypothesis of the primeval atom is the antithesis of the supernatural creation of the world," he insisted.[7] As far as he was concerned, nothing could be more divorced from creationism than the spontaneous workings of mathematical law. Believer though he was, no reference to his God seemed necessary to explain the Big Bang.

Almost a century later, this same conviction remains ironclad among many practitioners of cosmology, the scientific discipline that Lemaître helped to found. "The laws of gravity and quantum theory allow universes to appear spontaneously from nothing," wrote Stephen Hawking, the world's most famous cosmologist, in 2010.[8] And unlike Lemaître, Hawking set aside no room for faith to travel next to science. The prevailing assumption is now that God has been dispensed with: science alone can give a freestanding explanation of the universe's birth.

But this is a statement of faith, a mere conjecture; it is no more probable than any of the church creeds it meant to displace. Lemaître never really answered Hoyle's theological concerns. He simply waved them away, as public figures like Hawking have been doing ever since. The confounding moment of genesis remains stubbornly in place, and the question lingers: who or what set that first process of expansion in motion? For at the starting point of time, the universe is supposed to have been shriveled up within the smallest describable stretch of space, its dimensions contained within lengths of 1.6×10^{-35} meters (drastically less than the size of a proton). In the spaces bound by this minute distance, known as a "Planck length," the laws of physics so far elude our most advanced understanding, pitting quantum and relativity theories against each other in a confounding paradox. No one can yet describe the nature of things in such a tiny span, where gravity and time defy definition. Quantum waves of pure possibility would have had to convulse against each other without any restraint or stable boundary, rippling in a formless void. What spirit brooded over the face of that deep?

Seething with far more than white heat, trembling with an energy that could burst open the gates of being once untethered, the singularity tossed and roiled with nearly limitless potential. But it was *only* potential: matter itself could not take definite shape. Instead of particles there was a throbbing surge of possibilities, a hailstorm of all that might yet be. And yet from out of that miniscule rift, puncturing the monotony of infinite nothingness, there emerged all the forces now known in nature, rippling between the very first quantum objects. And no such object is really complete without someone to observe it.

So the usual story goes. It begins not with a collection of things so much as a tangle of probabilities, an unresolved array of chances waiting to become real. Among those chances, the likelihood of a universe like the one that emerged—perfectly balanced between matter and energy, calibrated to hang together in three dimensions of space and one of time—was vanishingly small.[9] "The probability wave meant a tendency for something," wrote Werner Heisenberg in 1958. "It was a quantitative version of the old concept of *potentia* [potentiality] in Aristotelian philosophy. It introduced . . . a strange kind of physical reality, just in the middle between possibility and reality."[10] This middle ground between something and nothing, between being and nonbeing, is where the universe would have stayed—had it not been seen.

Under normal conditions, a physicist might point out that "observation" doesn't have to imply "consciousness": in a technical term of art, scientists sometimes say that particles can "observe" and "measure" one another whenever they interact. A photon that meets with a moving object will become "entangled" with the object, so that information about the photon and the object becomes bound up reciprocally together: measure the object, and you know about the photon as well. Sufficient amounts of this "decoherence" could gradually interweave the whole universe together, meaning that all the possible outcomes of our various observations will be consistent with one another. But only one thing has the power to determine which one of those outcomes prevails among those that are possible: for matter to become finally definite, it must interact with a mind.[11]

And what was there to "interact" with the universe at the time-less moment of its genesis? What could it collide or meet with in those flashes of Planck time when its possibilities were first unfolded? And even if some rival theory should gain currency over Lemaître's, if the universe starts to look eternal or "steady-state," without a beginning in time, the problem of its resolution from possibility into reality will remain. "Universe" means "everything": from a simply material standpoint, there is nothing outside of it, nothing else to settle its tremors of potential into one real timeline. The wave function of one particle may get threaded together with that of another particle. But the wave function of an entire universe must meet with something *other* than the universe—something outside time and space itself.[12]

Einstein was right: beyond the day with no yesterday there yawns a limitless expanse of eternity without matter. And in that world beyond the world, outside the scope of every material object, there must be *something*—or else the universe could never have come into being. Even if the standard model of cosmology should be radically altered in the future—and the possibility is not a remote one—these basic philosophical questions will still demand answers. Nothing comes from nothing, as Aristotle knew: matter cannot set itself in motion. No timeline can unfold, no physical possibility can be made real, until it makes contact with something other than itself. If the sum total of all matter and energy is to meet with anything, it must meet with a mind beyond matter. For the world to be, it had to be seen.

Fracture

"For the earth was formless and void" until "God saw that it was good." The mind that turned its gaze upon the universe was light itself, "the true light that lightens all things." This story, ancient and all but discarded, now comes thundering back to life. It is almost an exact portrait of the world that science describes.

After all, if *no* mind ever looked with favor on the universe, it would have to remain a writhing coil of alternatives and contradictions. This is in fact the story that many devoted materialists now prefer to tell: no single cosmos developed from among the many quantum possibilities. Instead, the story goes that whatever could happen, did. The wave function itself describes reality; every solution to it is true, and every possibility it describes branches outward into existence like a fissure shivering and splitting its way over a frozen pond.

From our isolated perch on a single strand of this enormous cosmic web, we're told, we can only see one version of ourselves. Our single branch of consciousness snakes its way through an endless decision tree. But outside the edges of our vision there is a multiverse: "all the worlds are there, even the ones in which everything goes wrong and all the statistical laws break down," wrote theoretical physicist Bryce DeWitt, explaining this "many-worlds interpretation" of quantum physics.[13]

It began as a lonely suggestion by an embattled graduate student, Hugh Everett III. As a PhD student at Princeton, Everett argued that every possible solution to the wave equation really does hold true at once.[14] Schrödinger had already raised the possibility when he posed an impish question that would remain attached to his name forever: if a cat sits inside a box with a device poised to release poison gas upon the radioactive decay of one microscopic particle, will the quantum superposition of the particle eventually leave the cat both alive and dead? Everett's answer was *yes*: every chain reaction goes cascading outward along multiple timelines in every possible direction.[15]

In his day, Everett was laughed to scorn.[16] But he has cast a long shadow: multiverse proposals are now proliferating as swiftly as the alternate universes they depict, bubbling up one after another beyond the reaches of observation. Some of these proposals depend on Stephen Hawking's string theory, passing a quantum wave packet through a multidimensional landscape of energy vibrations so that it settles into those grooves most favorable for creating a new universe.[17] Others resurrect Everett's mathematics to split the timeline of our world at every moment into a dizzying array of multiple realities.[18] But in one way or another, all these theories hold out the promise—to those who find it attractive—of a universe that answers to no master. For those troubled by the thought of a divine eye opening upon the tender infant cosmos, multiverse theory offers a story of automatic genesis. No creator need apply.[19]

182 LIGHT OF THE MIND, LIGHT OF THE WORLD

But it won't wash. The problem with the multiverse isn't first and foremost a theological one. The problem is that it fails as physics. It makes a nonsense out of even the most basic claims about the universe that scientists are trying to put forward to begin with: "How can we say that there is a 50% probability of outcome A (which we can verify experimentally) if the truth is that all outcomes always happen (in some world or other) with 100% probability?" asks science writer Philip Ball.[20] Even more outlandish proposals—that we live in a simulation, or a hallucination, for instance—suffer from the same problem. If they are true, how can we trust any of our reasoning about anything? For that matter, how can "we" say anything at all if some doppelgänger version of us is always saying the opposite in some equally valid reality—if you both wake up living each morning and die unseen of a stroke every night? A man who answers both "yes" and "no" to every question isn't saying everything: he's saying nothing at all. A woman whose every choice fractures her into two women isn't living in a multiverse: she's dissolving into nothing.

Ancient Light

These efforts to erase consciousness from the origins of existence now amount to giving up on rational thought altogether, letting the very structure of language and mathematics crumble into dust. A steep price to pay for nothing in return. The elaborate edifice of the automatic universe is collapsing under the weight of its own contradictions like some old stone temple rocked by an earthquake. It's a venerable structure; at one time it even served us well. But now we are sprinting from one end of the tottering sanctuary to another, trying uselessly to plug the holes in its walls and hold up its teetering pillars. The strain of the effort is contorting our thought beyond all recognition. And for all the ingenuity spent on this doomed endeavor, we have not even answered the question with which we began: how can anything, or even the possibility of anything, come from nothing? If a quantum wave packet really did generate a multiverse, or an alien race wrote us into a simulation, where did the wave packet or the alien race come from? How can even abstract physical laws exist without a mind to lay them down?[21]

The structure of reason itself is compelling us now toward a new kind of discovery altogether. At the limit of all physical things—in the last regions where distance has any meaning, behind the thinnest instants of primeval time—there is a barrier. Beyond it, there is no matter. No thought that tries to cross that barrier can carry with it any mathematical models or physical shapes, no clothes or crutches to grasp on to. The mind must go there naked, as it was made, and meet with the mirror image of itself. For on the other side of that barrier is not void but fullness, the limitless soil of being, the golden womb from whose substance the world was made. And it is not water, as Thales imagined, nor prime matter, as the alchemists thought, nor particles nor quantum fields. The world is made at last from none of those. It is made from the deep that cries out to deep, from the light that meets each mind at the threshold of existence with a servant's silent welcome and a king's stately proclamation, bound up together in one saying: "I AM."

Any voyager brave enough to meet this moment will feel a sense of recognition: *it was you, all along.* The human mind has gone on a long journey into unfamiliar territory, through a wilderness populated by strange gods and deceitful ghosts. In the desert, the mind forgot itself, drawing away from the world until planets and particles alike were mere objects drained of all life. It was a journey we had to make, a retreat to a high and lonely mountain where we could look down at the universe and master it. But something wonderful has happened on the mountaintop: the same science that removed us from the world has now forced us back into it, and our wandering has become our homecoming. At the extremes of physics, the human mind can be led once again to recognize itself, to realize that it is not an illusion or an accident but a shadow cast throughout the universe by the wisdom that set the stars in order.

Once this is acknowledged, the heavens and the earth come roaring back to life. The improbable composure of our universe amid innumerable possibilities becomes a matter of choice: ours is the world selected among all possible worlds to sustain the kind of life that we are. Science, too, makes sense again: all the exiled ghosts and small gods take their place as ministers before the one great mind that gave them shape. Energy and particles can be spoken of, because they were seen. Out of a formless tangle of potential time-lines, one eternal instant of observation set the wave function of the universe collapsing into a single series of events.

At this point, most models say, things began to cool and take shape. At first radiation and matter were intertwined in a dense plasma, meaning that no light could escape to meet with present-day measuring devices. But after about 380,000 years, as measured from our human point of view, atoms formed that did not latch on to photons, allowing the latter to travel freely through space. The universe then became "transparent," meaning that for the first time, light could move across huge distances and make an impact at last on our most refined instruments. This is the first moment of recordable time, the instant when light broke free of matter and made the universe possible for humans to see.[22]

Everything before this point is separated from us by an unbridgeable divide. What happened before light and matter

separated is one unmeasurable sequence of events, accessible only
by theory and indirect speculation. What happened after the light
broke free is visible to us through telescopic and satellite measurements. In other words: "God divided the light from the darkness,"
the first period of creation was brought to its close, and a new one
began: "there was evening and morning, the first day" (Genesis 1:5).

Turn on an old-fashioned television and set it to a channel where
nothing is playing. Amid the static picked up by the antenna, there
are stray photons from the first moment when the light of the universe could be seen by human eyes.[23] "Cosmic microwave background radiation," the ambient wash of ancient light from the farthest reaches of recordable time, hangs in every region of the air as
a relic of creation. Just before his death in 1966, Georges Lemaître
heard news of its discovery. He died amid the unseen glint of light
scattered from the world's beginning, photons like angels rushing
through the air from the universe's first day to the priest's last.
Perhaps in that moment the wall between human and divine things
was not quite so thick as he imagined.[24]

Works and Days

Perhaps, in fact, the two worlds are layered on top of one
another—nature and supernature, science and theology, fused as one
in each moment of human experience. The events of the physical
world are already teeming with more than material life; even the
words we use to describe space and time are aglow with the secret
light of the mind. "Energy" is a power we experience coursing
through the objects around us, even though it has no tangible substance to speak of. "Time" is the unfolding of motion and change
around the still point of an observer's position; *past* and *future* are
words with no meaning unless they extend outward from their

meeting point at a present defined by a watching mind. Even "particles" and "waves" are units for measuring and organizing experience—empty vessels to be filled with the qualities and quantities we measure. From the first syllable of time, the world has been a language—spoken, and meant to be heard.

Skeptics may protest that if 380,000 years really did elapse between the big bang and the separation of light from matter, then that is a far cry from the "one day" described in Genesis. But this is exactly the point. These are two totally distinct ways of speaking—the modern scientific way and the ancient spiritual way—which may nevertheless describe exactly the same thing. It would be doing the author of Genesis a disservice to insist that he must have imagined time unfolding in human terms before there were any humans to measure it. That is a modern conceit, a product of the kind of thinking that has replaced living experience with an imaginary picture of mechanical motion.

Ever since "truth" became a matter of mindless objects moving through space, ancient authors have been held to a standard imposed by the heirs of Galileo. The standard can be summarized as follows: physical truth is the only kind of truth there is. Unless our forebears described the world in numerically "accurate" and physically "objective" terms, they were either lying or simply mistaken.

This has been a disaster for our understanding of basically any author who wrote before the scientific revolution. But it has left us particularly unprepared to hear what the authors of the Bible are saying and how they are saying it. For even when defenders of scripture leap to prove the authors of the Bible correct, they often do so in terms of exclusively material truth: actually, they argue, the Bible gets the "facts of the case" right and the scientists have it wrong. It really was twenty-four hours between each day of creation, not thousands or billions of years.

This is certainly a position that some august interpreters took before the scientific revolution.[25] But it was not the *only* acceptable position for Christian or Jewish believers.[26] It is a modern conceit that no reading of Genesis can be faithful unless it describes a young earth—a position which implies that *unless* the Bible means "twenty-four hours" every time it says the word *day*, it must be wrong.

This is giving the game away before it starts. It is a trap to get caught in a competition over mathematics, trying to prove that the days of Genesis were exactly those same intervals of time that human beings measure by the appearance of the sun from Earth. It amounts to accepting from the outset that time and motion are invariant units of measurement, applied to a world of matter which goes on rolling along just the same whether we are there to observe it or not.

If I said that the earliest human beings lived "beneath the wisp'ring roof / of leaves and trembled blossoms," no one would think to ask me where they got the tools to build a house. I would be using images from our present experience (roofs, architecture) as a way of explaining to modern readers what it was like before any such things existed, when mankind took shelter under trees and bushes. Just so, if I were trying to tell a story about the world before the sun and the moon to people who had only ever seen time passing by the light of those two heavenly bodies, a "day" might be the nearest thing in my listeners' experience to the interval of time I was trying to convey.

This is not some modern innovation to evade a choice between science and religion. "Our ordinary days have no evening but by the setting, and no morning but by the rising, of the sun," wrote St. Augustine in the fifth century AD; "but the first three days of all were passed without sun, since it is reported to have been made on

the fourth day." And so "evening" and "morning," Augustine reasoned, must mean something other than the setting and rising of
the sun: "What kind of light it was then, and by what periodic movement it made evening and morning, is beyond the reach of our
senses." And if the aftereffects of that creative ecstasy are now beginning to reach our senses, should we be surprised that they seem to
come to us from a great distance of time, as measured in our human
years?[27] The days that unfolded before God's sight alone were the
stages of creation as it was being prepared for the eyes of his creatures. Once those eyes opened, the whole world was destined to take
on new shape and character.

Welcome Home

No great surprise, then, that the history of the universe has unfurled
behind us like the rich carpet that welcomes a prince into his palace.
The billions and billions of years that we have measured before our
existence are years of *our* time, as it appears to us from our stand-
point within the universe. In fact, it may even be that those yawning
ages only opened up once we measured them. They could not have
been otherwise—we do not have the power to remold time at will.
But time as we experience it may require *us* to make it real. And this
is perhaps the most suggestive possibility raised by quantum physics:
it seems possible that our measurements may reach *backward* and
help define the past.

That was the insight of John Wheeler, the Princeton physicist
who pointed out that a measurement taken today can affect the
trajectory a photon took *yesterday*. "There is a sense in which the
polarization of those photons, already on their way, 'was' brought
into being by the disposition of the polarization analyzers that the
observer has yet to make and will make," Wheeler wrote. "We are

not only spectators. We are participators. In some strange sense, this is a participatory universe."[28] Perhaps the light that travels to us from distant stars, sailing around the curves made in spacetime by the heavenly bodies, only really charts out its trajectory once it meets our measuring devices.

But the arrival of light from distant galaxies after long years of voyaging is part of how we look back into the past of the universe. So, the exact path of light, determined when it reaches us, helps shape the past that elapsed long before we existed. It might not be too much to say that we bring this distant past into being by giving galaxies the definite history they would lack if no one had observed them. "Observership draws the past more firmly into existence," writes Stephen Hawking's protégé Thomas Hertog.[29] The vast age of the universe does not dwarf our existence: our existence *confers* age onto the universe, bestowing upon it the venerable honor of becoming our ancient home.

The cosmological vision outlined by Lemaître, Edwin Hubble, and their followers is currently in the midst of a drastic re-evaluation. Some of it may have to do with the indispensable role of human observation. As images come streaming in from NASA's James Webb Space Telescope, anomalies in time are causing scientists to question the rate of the universe's expansion and the age of galaxies that seem far older than their distance from earth would imply.[30] Proposals have been floated to increase the age of the universe twofold, but even that wouldn't resolve all the complexities and inconsistencies that are currently wrinkling the outlines of deep time.[31] Perhaps the answer lies in us: as these galaxies reveal themselves to human observation for the first time, new portions of their history are being called forth into existence by the appointed regents of the universe. In the days of creation, the ages of God's time, they were made ready for us—bathed in the light of the world. Now they meet with the light of the human mind and take their final form, like servants putting on their livery to meet the heir to their castle's throne.

"Within you, O my mind, I measure times," wrote Augustine.[32] Only time, indeed, will tell how radically the standard model of cosmology will have to be revised. It may be Wheeler was on to something, and our participation is involved in the creation of new galactic histories. Or it may be that other answers suggest themselves as new measurements emerge. But no measurement can make any sense at this stage unless we confess that it comes in part from us—that the world is not indifferent to us, is not a wasteland in which we happen to occupy one inconsequential backwater. In any model of cosmology that can possibly carry conviction, the world needs a watching mind to make it whole. "My delight is in the sons of men," declares God's Wisdom, who danced with him "at the creation of the earth" (Proverbs 8:22–31). From the first word of creation, from the day with no yesterday when time began, the whole bejeweled kingdom of heaven and earth was made to welcome us. It has been holding its breath until we arrive.

CHAPTER 10

The Word

Evermore
From His store
New worlds rise up to adore.
—Joachim Neander, hymn: "All My Hope on God Is
Founded" ("Meine Hoffnung stehet feste")

F ew mysteries run deeper than our participation in the universe. It is so fearsome to contemplate that we will do anything to ignore it or disguise it. But we must face up to it now, or risk squandering our birthright.

For centuries we have been locked in a civilization-wide struggle to understand creation without reference to any human involvement. Recovering that involvement now can feel like an invitation to shamanism and crankery. In fact, the discovery of quantum mechanics did inspire a small platoon of wide-eyed pseudo-yogis and self-proclaimed Zen masters, eager to announce the supposedly scientific discovery that "the world is what you make of it."[1]

Objectivism—the deadening philosophy that reduces the world to material objects—has proven so inadequate for describing reality that it has inspired a desperate hunger for some alternative outlook, some way of seeing that will rescue the truths of human experience and emotion. But since the facts discovered by science are inescapable, it came to seem as if the only obvious refuge from them was in outright denial of truth altogether.

This helps explain not only the popularity of occult conjurings like neo-astrology and crystal therapy, but also the whole postmodern conceit summed up by the French philosopher Michel Foucault: "truth is a thing of this world." If reality "is produced only by virtue of multiple forms of constraint," as Foucault put it, then there's no threat that science will erase mankind: mankind and science alike are social constructs enforced by power.[2] Both can be remade with sufficient application of the will.

Much like new-age spiritualists, postmodernists hope to wriggle free from the iron grip of objectivism by grasping onto its opposite extreme: pure subjectivism, according to which the mind is altogether sovereign over reality. If reality is exclusively a product of our thought—if truth can be altered through meditation, or ayahuasca, or the violent enforcement of political euphemisms—then no one needs to trouble himself with "objective facts."

But this subjectivism is a dead end. All it does is further discredit the possibility that human experience may be an essential feature of reality. By putting forward impossible claims on behalf of the human mind, subjectivists make it look impotent when it inevitably fails to deliver. Obviously, we *can't* reshape existence just by thinking, or alter material facts by wishing them away, or turn men into women by applying different pronouns to them. Since claims that we can are plainly false, they only serve to make objectivism look good by comparison, reinforcing the impression that science alone offers a secure route to truth.[3]

In the last analysis, objectivism and subjectivism are two faces of the same superstition: equally extreme and equally impossible to credit. The world is neither a dead object nor a construct of the human subject. It is a *relationship* between subject and object, a meeting between mind and matter. When an author writes a book, he infuses paper and ink with the characters and events he has imagined. But no author throws his finished masterpiece into a vault and calls it a success. The book needs readers; in some sense it is not really a book until the life that teemed behind the author's eyes is transferred from the page into another mind.

If a single copy of *Fathers and Sons* is burned, no one says that Turgenev's novel has ceased to exist. The only way it could ever vanish completely is if every copy were destroyed and every mention of it forbidden, if there were no one left on earth who even remembered reading it. In this sense the book is not the paper, or the ink, or even the author, or the reader: it is what comes into being in the space between them.

The Great Communion

The real suggestion of quantum physics is not that the world is what we make of it. It's that the world comes into being among and between us. Whatever *we* mean when we speak of things—from the stars in the firmament to the smallest particles of light—we are always talking about the way things affect *us*. We are talking about what is born of the congress between matter and mind, as a story comes to life in the relationship among the author, the reader, and the page.

The dizzying fear that rushes in at this discovery is a fear of relativism. If the world is dependent on the way we perceive it, how can truth be anything more than a figment of our imaginations? But this is a fallacy, generated in no small part by the false choice between objectivism and subjectivism. We mistakenly imagine that if something is not a purely physical fact, it must be a purely arbitrary mental construct. Those are not the only two options.

Consider again John Wheeler's hypothesis about the trajectory of light from space. Wheeler thought that human measurement could help determine exactly which of many possible paths a photon traveled from a distant star to earth. Before the light registered on a measuring device, its motion was not described by any one definite line. Once it showed up on earth, its particular path had been determined: the observer in the present reaches back into the past to make it real.

But what Wheeler did *not* say was that each observer could choose his own path for the light to have taken, or that once the path had been determined, it could be revised. Observations, once made, are fixed: they are not private property or toys to be played with. The world, once observed by even a single mind, belongs to us all.

Or else, by way of analogy, consider the rainbow. Its array of colors has no physical substance; the strips of red and indigo that arch across the sky are nowhere to be found in the moisture that hangs in the air or the photons that come streaming from the sun. The most that physics can say about those photons as they scatter through the raindrops is that their energies are associated with certain wavelengths, and those wavelengths register in the cones of the eye. But wavelengths are not colors, and no quantity of energy has the glow of yellow or the flush of green.

Only the mind can interpret the impact of light on the eye as the quality we call color—not the material interactions that give rise to it, but the actual color, rich and warm and alive. What we call the *rainbow*, the thing itself, hovers in the space between the viewer and the air around him; it bursts into existence when the light of the mind meets the light of the world.[4]

Still, no one can refuse to see a rainbow when the light strikes him just right, nor can I decide that there is a rainbow in my room on a dark morning. Anyone who said he had the power to create rainbows simply by thinking about them would be either lying or deluded. The fact that we bring certain things into existence by perceiving them doesn't mean we get to choose how and where they appear, any more than I can make the books on my desk vanish by closing my eyes. We can compare and discuss the things that we see—they are not simply private hallucinations but shared creations, things we receive together.

In just the same way, no one can wake up tomorrow and decree that photons don't exist, or manifest electric currents by sending thought waves out of his fingertips into a telephone pole across the street. Yet at their most basic level, in their raw state of existence, even photons and electrons are born in the communion between mind and matter. There is no such thing as a photon in total isolation, floating completely on its own—at least, it is not the kind of thing we have in mind when we speak of an "object." There is only potential, a range of possible places and states, a sentence waiting to be read. To take shape and form, to put on full and final being, even particles must be seen.

In the world of our daily experience, these strangest of quantum effects are constrained by the very structure of things. The uncountable throngs of particles that make up chairs and tables are already embedded in a network of relationships, bound together in every direction by their contact with one another, so that each of their myriad possibilities is resolved. They become real by virtue of being an integrated whole: once the protons and electrons in my desk are locked together in relationship, I can say with confidence that the table itself will not be on the other side of the room the next time I look at it.

But in the behavior of the smallest things there is a whisper of grander truths. If I were to peer ever more closely at the table, to study its parts in isolation, I would eventually reach a point at which the disassociated fragments became small enough that their location and movement depended again on my measurements. The more things break down, the more separate they become from one another, the *less real* they grow, until they pass out of the range of human observation altogether and become figments of pure abstraction. It is in relationship—with each other first, and then finally with us—that the things in the world take on their full character.

We don't learn the true nature of things by dissolving them into their component parts. We may learn more about their patterns of behavior that way. But we haven't "looked beyond" or "underneath" their false exterior to some more "real" mathematical truth. We haven't looked "beyond" anything at all. At every level, the things we see around us are products of the interchange between us and the world. They are born like children from our minds, implanted in us by the mind that made them possible.

Our human experience of them—the motion of the stars across the sky, the cascade of nebular dust as it reveals itself to our telescopes, the brilliance of the sunset or the rainbow—these things that we can experience and speak about together are the fullness of what we mean by "the world." We will never get "beyond" it. We will never need to: it will continue to reveal new richness, new hidden chambers, new corners and vistas, as we enter into new relationships with creation.

To our ancestors the stars appeared as points of light; to us they appear also as blazes of fire against the backdrop of space. But without any mind to see them they would look like nothing at all: their matter is clothed by the mind in a thousand garments, rich every night with new beauty, capable of being seen from as many points of view as there are minds to perceive them. Every consciousness that is born, every eye that opens, calls forth from the stars another form of glory, draws forth another appearance from them like water from a cistern. At the dawn of time they were charged with more potential than we could exhaust in a thousand lifetimes; whenever we see them, we fulfill a promise that was made in them when they were born. Our office and our delight are to look at the world in all the distinctive ways that humans can, so that in communion with us it can come to fruition.

It is likely that other animals have their own modes of vision and understanding. It may even be that there are forms of consciousness active in the wider universe that we cannot yet reach or understand.[5] We don't know—what we do know is that our way of seeing *does* something to the world: from the most rudimentary granules of existence to the full sweep and scope of space and time, things are different because we are here.

Word and World

Human consciousness doesn't just sit dumbly as a barrage of sense perceptions gushes into the mind like a jet stream. We never just receive a wash of undifferentiated color and light, or a gale wind of unconnected particles: we gather our perceptions into objects with form and order. When we see things, our minds give shape to them. In the most profound sense, we give things *names*.

It is said that after creating the first man, "God molded every living beast of the field from the soil, and every bird of the air, and brought them to the man, to see what he would call them. And whatever the man called each living soul, that was its name" (Genesis 2:19). More was bestowed on the animals that day than just sounds. The names we form with our mouths or write on a page are the tip of an iceberg whose unseen mass hovers in the silent depths of the soul. Spoken words are the final product of a process that begins, as the medieval theologian Thomas Aquinas knew, in the mind and the heart. The intellect, wrote Aquinas, "receives a light" from the *species* of each thing it perceives—that is, we see the kind of thing it is, in the way it appears to us.

A man confronted with an eagle is coming face-to-face not just with a single bird but with the representative of a type. To call it an "eagle" or a "bird"—to call it a *nesher* or *oiseau*, to call it *anything*—is to place it in a clan with other eagles and other birds, to assign it a place among the ranks and battalions of winged predators, and ultimately to enlist it in the raucous army of created beings whose ranks teem across the whole earth. Like a general inspecting his troops, man sets the world in order by understanding it, by coming to know the nature of things.

The words we say and write are outward symbols of an inward grace, an intimate connection between us and the things we see.

The eagle is more to us than just an individual eagle: whatever the bird may think of himself, the soul of man stamps onto him the sigil of a race and a tribe. There is a reason why sports teams pick wildlife as mascots to rally behind, and why the ancient tribal armies of Europe marched with totems carved in the shape of beasts. There is a reason, too, why the beasts in the enchanted woods of fairy tales can speak with humans, and why even the old gods walked the earth with favored animals prowling at their feet or hovering by their shoulders—Jupiter with his eagle, Athena with her owl, Dionysus with his leopards. Pagans, too, understood that the creatures of the earth and air were more than biological curiosities. Each had a peculiar and distinctive nobility that was drawn out of it by human perception, like the horses of Achilles that Hera taught to speak. The soul of man gives a soul to the world.[6]

We do this naturally, before our mouths ever form a single syllable, as part of our experience of creation. God did not present isolated creatures to Adam but classes and types of animals, each after its kind: "the birds of the air and the beasts of the sea." The world as presented to us already comes infused with order and meaning—Aquinas says that the light which floods the soul is there "when the object is formed, and remains with the object formed."[7] When we see the world, we draw out of it an order that was implicit in its making. It becomes real in our sight.

At some point in the development of life on earth there must have appeared, for the first time, a creature who could use language in this deepest and most primary sense. In that instant someone (and it is at exactly this moment in prehistory that it would no longer be appropriate to say "something") opened his eyes upon a world not merely of sense perception but of meaning and form. "This creature was truly different from all other creatures," writes the apologist G. K. Chesterton, "because he was a creator as well as a creature."[8]

This is not the place to haggle about when and how it happened; it is for others to puzzle over fossilized skulls. The point is that at some moment a creature was born who experienced not merely a succession of unconnected sights and sounds but an interconnected web of forms and species, a world of things living and dead connected through time. In that moment the sun rose as never before on an earth that could be understood by someone within it; the days and nights took on a rhythm that could be felt, and one by one the animals received their names. It is not too much to say that even past, present, and future took on a new shape, and only then was the work of creation complete.[9]

It might be no accident that Homer, our oldest surviving Greek poet, represented certain heroes as having been touched by the gods and blessed with the authority to rule through reason and debate in counsel. Reason and debate having newly come into existence, a different kind of creature altogether would have emerged, the rightful lord of the world. He was connected not only to earth but to heaven, capable of bridging the gap between flesh and spirit. The moment such a creature appeared, the world was made new—for God had made man.

A Place to Stand

"Plant a seed in the earth," wrote the Austrian visionary Rudolf Steiner. "It puts forth roots and stem, it unfolds into leaves and blossoms. Set the plant before yourselves. It connects itself, in your minds, with a definite concept."[10] Like the tendrils of a sapling unfurling steadily from their source, the world grows and lives when

presented to the mind of man. As a species, humanity calls forth
something new from the world, a blossom it would not bear without
us. And like a mustard seed, the world is already loaded full of forms
and meaning, ready to bear fruit. Human experience is not an accidental byproduct of the universe but its consummation.

In this sense it was very truly said that "in the beginning was
the Word," and that "without the Word not one thing was made that
was made" (John 1:1–3). This was an answer, for those who could
hear it, to the question that gripped the anguished heart of Plato
when poison stopped the heart of Socrates. How could the noblest
mind in the world be encased in a body that could decay, a nervous
system that could seize up and grind to a halt in the grip of something so earthbound as the juice of a flowering plant? How could
eyes made of all-too-supple fluid gaze on untouchable beauty? How
could the realm of things beyond the senses be known through the
touch of flesh?

In one form or another, these questions have tormented mankind for as long as anyone can remember. By the first century BC,
few would have looked for an answer to come from a restive but
ultimately insignificant corner of the Roman Empire, where a ragged
assortment of mismatched companions huddled together around
their rabbi in observance of a high holy day. But that was the night
when another and still greater soul took leave of his friends. He too
would soon be pronounced an enemy of the state and a traitor to his
people; he too would be put unjustly to death.

Before the eyes of the beleaguered Israelites, Jesus of Nazareth
had done things that should have been impossible. He had sum-
moned fish from the sea and bread from the air, defying the cruel
laws of scarcity that seemed to govern the earth. Some of his
admirers had dared to hope that he would lead them to an

improbable triumph over the worldly powers whose dominion was so punishingly evident in the callous abuses they had suffered under Roman legionaries and tax collectors. Now, apparently, it would all come to nothing: inexplicably, the man who could walk on water was going to hand himself over without a fight to the authorities who wanted him dead. His most devoted allies clustered around him; bewildered and hurt though they were, they had nowhere else to go.

One of them, "the disciple whom Jesus loved," spent the night "reclining with his head against Jesus' Chest" (John 13:23). And maybe it was in that evening, while listening to the steady heartbeat of God, that John conceived what would become the opening to his Gospel. Certainly he would have contemplated how it could be that such a fragile muscle, thudding dully within the rib cage of a man condemned to die, could be throbbing with the lifeblood of the source of all life.

And when the night of torture was over, as reports went out abroad that the same man had stood up and walked out of his own tomb, maybe John remembered the sound of that beating heart. If the life of all life really could be sustained by blood and sinew, if the mind that commanded the mountains to rise had occupied a skull of bone—and if, when that eternal soul consented to subject itself to the awful material laws of death and decay, it was the matter that had surrendered and death itself that had made concessions—then perhaps flesh was *always*, from the foundation of the world, meant to serve as a vehicle for the spirit.

In fact, it was never otherwise. In fact, wisdom danced with God when he marked out the boundaries of the sea, delighting in the sons of men. In fact, the human way of seeing the world and speaking about it, the form and structure that human beings discern through the medium of their senses, was never an accident or a triviality. For the *logos*—the "firstborn of creation," the word that is before any written letter or spoken sound, the reason and meaning that cannot be altered or gainsaid—was infused into matter from the moment it was created. There was never just "stuff," mere objects, matter without mind. Each and every thing that was made was invested with more than material properties, suffused with "life, and the life was the light of all mankind." Man, in God's image, draws forth the true character of His creation, a world not only of objects in space but of virtue and desire, longing and loveliness. We are here to give names to what the Word has made.

There is now a powerful impulse in us to try and extract ourselves from the world. Possibly there always has been. This may be part of what it means "to become as gods, knowing good and evil." The most ancient temptation is to stand outside the world and judge it with dispassion, weighing the cosmos in our hands like the solemn deities of Olympus, as if "good and evil" were matters not of personal investment but impartial scientific study.

Archimedes, the Greek mathematician whose lordly insights gave him a sovereign indifference even to kings and executioners, is supposed to have said, "Give me a place to stand, and with a lever I will move the whole world."[11] This longing—to gain purchase on things so that they don't affect us, to extract the human term from our equations—runs very deep. In the scientific age—in our age—it has become a messianic world project, an impulse to tabulate and commodify the universe. "You shall become as gods, knowing good and evil"—knowing them as figures to be counted, not felt; knowing dejection and despair as neurochemical imbalances; knowing violence and crime as a mere misallocation of social resources; knowing the body itself as "a toy or a pet" to be played with and carved into pieces.[12]

204 LIGHT OF THE MIND, LIGHT OF THE WORLD

But the great secret is that this is not how such things are known. Magnificent though the achievements of science have been and will doubtless continue to be, they came from the beginning with a whisper of that old temptation to judge the world, this time in terms of numerical weights and measures. But even those dependable figures are framed within the human mind, shaped and structured like everything else by the communion of matter with consciousness. There is no point from which to move the world, because the world is made when we are in it. Take us out and it becomes a dead sculpture, upon which we gaze with the empty stare of idol worshippers.

Death and Life

This lightless idol of a world will demand what all idols demand: sacrifice. Like the somber carving of the god Moloch, who waited dead-eyed in the killing fields for offerings of infant flesh, the universe without its human regents becomes a bottomless pit, provoking in all who approach it a fanatical hunger for annihilation. The sinister logic of idolatry is to promise life and deliver death; this diabolical bargain is now taking place at a cosmic level. The universe itself is being turned into one giant stone altar, on which penitents are rushing to obliterate themselves.

"Humanity is actually acting as a destructive force on Earth, and we are losing control of it through pollution and its effects on ourselves and the planet," we read in the *Antinatalism Argument Guide*: "The only way to avoid it is to have no children."[13] Antinatalism—the conviction that the world and the environment would be better off without any more humans—is a creed whose passionate evangelists imagine a universe emptied out of its most monstrous inhabitants.[14] "Look at what we did to this planet," says Les Knight, founder of the

Voluntary Human Extinction movement. "We're not a good species." Others, in awe of the dispassion and the sheer power exhibited by machines, hope that mankind will be superseded by its sleeker, smarter children: "people of flesh will upload into software the contents and processes of their minds," predicts Martine Rothblatt: "new digital people can be produced by combining some of our mindware with some of our partner's mindware."[15]

Whether through physical or spiritual castration, whether in the name of nature's unforgiving gods or in admiration of cold steel, men of tremendous power and determination are preparing now to fling themselves and their disciples into the dark embrace of death. What they imagine in that great beyond is liberation and majesty, a glittering paradise where human limitations vanish like a passing nightmare at the dawn of day. But it may well be that beyond the boundaries of the human body is not a gleaming expanse but a colorless void.

"The world, which seems / To lie before us like a land of dreams, / So various, so beautiful, so new, / Hath really neither joy, nor love, nor light." So wrote the poet Matthew Arnold at the prospect of material prosperity without the virtues of the spirit. Dazzling as our technological achievements may be, the world that they create will be a gray and silent one—will not even truly be a world—unless we are there to give it life.

This is why the light of the mind is not simply a philosophical curiosity or an interesting feature of theoretical physics. It is a matter of desperate urgency, the difference between apocalyptic horror and glorious revelation. Only return humanity to the center of things, and a future that now seems fraught with menace begins to quiver with signs of hope. We have hardly begun to imagine what the universe could look like if we recognized it as our home—what it would mean to colonize planets, to suffuse the world with nuclear power, to give sight to the blind, if the eyes that opened on the new world were human eyes.

The technology to do all this, too, is coming into being. If brought into the service of humanity, it could indeed deliver on all the extravagant hopes of its creators. But to serve humanity truly requires a knowledge and a love of man as he has been since the first moment he was lifted from the dust—not a mistake or an irrelevancy but the priest and shepherd of a more than material world. It is by knowing and understanding *that* world, with its physical and spiritual truths alike, that we bestow on it a final blessing.

The infinite ecstasy of creation, God's act that stands forever at the beginning of time, finds its answer again and again in the first words of every child. From the first instant when understanding makes its way from the heart to brighten infant eyes, the universe is awash again with the light of the mind, and all things are made new. All the rest of our acts, from breastfeeding to astrophysics, proceed from this first moment—old as humanity itself, refreshed again with every human birth.

The poet Coleridge wrote that man's literary creations take on life "when a human and intellectual life is transferred to them from the poet's own spirit."[16] As in poetry, so in architecture and engineering, so in physics, so in the simplest exchange between mother and child: every eye that opens on the world imprints it with "a human and intellectual life." And this is how our interventions in the world can become acts not of destruction, but of praise.

Scratched on the back of an inventory receipt from the third century AD, there survives—just barely—a form of ancient sheet music that records the oldest known melody to a Christian hymn. There is just the shred of a closing lyric, a snatch of otherworldly singing: "while we sing hymns to Father, Son, and Holy Ghost, let all the powers join the cry: 'amen, amen.'"[17] It is among the oldest themes of Christian worship that humanity has been charged with gathering up the praise of the whole material universe, from the creatures that lumber in the fields to the stars whose pulsing light almost gives the impression that they are sobbing for joy.

"The skies are a book; God's majesty is written on it," goes one ancient song: "Heaven's dome is a declaration of what his hands can do. Day pours forth speech to day, and night reveals knowledge to night. No, they have no speech, no utterance—but still! Their voice is heard." (Psalm 19:2–4). The theme resounds in hymn after hymn: "sing with all the people of God, and join in the hymn of all creation."[18]

Secrets that have been murmured for centuries in prayer and song are coming shockingly to life now in the field of physics. The natural world itself is testifying: for all that we have abused it, science itself is revealing how deeply the words of creation are etched into the cosmos and into the human heart. The silent voice of the days and nights, the code of DNA, the hidden mysteries yet to be unfolded from every grain of dust—all of this is a language to be read by none other than a human mind. In 1989, John Wheeler proposed that "every item of the physical world has at bottom—at a very deep bottom, in most instances, an immaterial source and explanation."[19] The root of existence is not objects but "information"—this idea of getting "'It' from 'Bit'" is a new way of expressing the very old insight that the world was made to be known.

"*Mine* is the sunlight, / *Mine* is the morning," goes another hymn, "Born of the One Light / Eden saw play." Every human eye that opens bears witness to "God's re-creation / Of the new day."[20] Though there is only one universe, it is remade as many times as there are humans born, seen and anointed newly from the stand-point of each indispensable soul. "Thou wholly communicatest thy-self to every soul in all kingdoms," wrote Thomas Traherne to his God, "That all thy saints might live in me, and I in them."[21] So too the apologist C. S. Lewis: "For every soul, seeing Him in her own way, doubtless communicates that unique vision to all the rest. That, says an old author, is why the Seraphim in Isaiah's vision are crying 'Holy, Holy, Holy' *to one another (Isaiah VI, 3)*."[22]

To bear witness to the universe and anoint it, to call forth from it the glories that were hidden in its most secret parts from the moment of its birth: that is the task of the human race, the living work of praise that will occupy us until the end of the world. The smallest particle that flits across our fields of vision, the grandest galaxy that wheels across the heavens, the fish of the sea and the birds of the air: all of them find their form and meaning in our sight. If we lose this knowledge—and we are in danger of losing it—we will lose the entire world. But if we can recover it, nothing is beyond our reach.

The universe is still waiting for its master's children to return from exile. We have wandered deep into the wilderness, desperate to escape our own selves and our own calling, desperate to reduce the world to mathematics or to machinery. But the hour is growing late. It is time to come back home to what we know to be our birth-right, to our role as stewards and humanizers of the universe. When we come—if we come—we will be greeted by that very universe with songs of praise that link our newest achievements to our oldest tradi-tions. And it will be our task again to gather up those praises so that all the powers join the cry: amen, amen.

An Invitation

We have a choice to face. It is a choice between two visions of the world. The first is well established and even canonical, a kind of dogma. Gradually, what began as a convenient fiction has become a hidebound absolutism: for the messianic prophets of our dangerous age, the world is an object to be dominated and humanity a primitive stain to be wiped from the face of a perfectly mechanized universe.

But cutting-edge science and inherited wisdom alike are revealing a totally different picture—a picture of a world that cannot be what it is without us, that is endowed with all its meaning and structure by our presence in it. A world spoken into being by the eternal Word and made complete by the words of his creatures, who bear his image on Earth.

That story—ancient but forever new—now stands as the indispensable, the desperately needed alternative to a savage and ultimately irrational new faith, a grim and monstrous new tower of Babel whose silhouette is already rising against the skyline. There has only ever been one proclamation that can bring such towers crashing to the ground and leave their builders in confusion: the proclamation that in the beginning, God created the heavens and the earth.

The Hungry Sheep

This being so, believers of a certain cast of mind may have found this book overly concerned with the details of physics and cosmology. If scientists will accuse me of bringing religion into science, theists will accuse me of bringing science into religion. Why even *try* to "reconcile" the truths of Genesis with the vainglorious fancies of prideful man? The Bible is clear, these critics will argue: the earth was made in six twenty-four-hour days—why muddy the waters by introducing arrogant theories invented by enemies of the faith, who set no store by the word of God and whose imaginings will vanish in a moment like the flower of the field?

"Who are the people that are greatly troubled by new systems of philosophy and infidelity which are constantly springing up?" asked Charles Spurgeon, the great Baptist theologian who is known across various denominations as the "Prince of Preachers." "Certain young folks say to me, 'O sir, I have read a new book, there is a great discovery made about development. Animals were not created separately, but grew out of one another by degrees of gradual improvement.' Go and ask your grandmother about it!" laughed Spurgeon.[1]

By this he meant that the people of God were not accountable to every sophist that came breathing the spirit of the age. He imagined the faithful old woman answering her grandson: "'I know whom I have believed, and am persuaded that he is able to keep that which I have committed unto him against that day.' You think her a simpleton, perhaps: she might far more properly think you the same." I am sure that many people, reading my speculations here about singularities and the ages before light was divided from darkness, will think similarly. Why complicate simple scriptural belief with the meddling of scientific speculation?

I can answer this question best with a story. When Albert Einstein was a boy, he loved the Bible. He was hungry from his earliest youth for an approach to life that took it seriously; in scripture he found the weight and purpose that he longed for. "I came—despite the fact that I was the son of entirely irreligious (Jewish) parents—to a deep religiosity," wrote Einstein. But his faith was promptly stifled at age twelve when, "through the reading of popular scientific books I soon reached the conviction that much in the stories of the Bible could not be true."[2]

Einstein was a man not just of brilliance but of decency and sincerity. A man like that would not have discarded his childhood beliefs just because he wanted to vaunt his own pride or deceive himself. Here was a boy straining honestly and vigorously to understand the world. What he was looking for from both science and the Bible was an explanation that had clarity, purpose, and internal consistency. And he felt that science could deliver truths that the Bible and the Talmud could not.

The fact that he would endorse science at the expense of religion reveals that he viewed each of them as routes to the *same kind* of truth. So, where they differed, how could he do otherwise than view them as competitors? Einstein grew up with the assumption that both the Bible and his physics textbooks were making physical claims about the world in literal terms. When he read about the earth being created in six days, he took this to be a false claim where science offered him the hope of a true measurement.

And if truth is only a matter of objects moving through space, who could blame Einstein for casting his Bible aside? He had weighed it in the balance and found it wanting. But when he did so, the balance itself—the very system of measurements and the criteria of truth he used—were those set in place by the overconfident ministers of the scientific revolution. In other words, he had already accepted an idea of the world as a collection of moving objects. When he found that the Bible did not describe the world that way, he looked for a book that did. And what he found was science textbooks.

This train of thought, I suspect, has been followed by generations and generations of men and women both before and after Einstein. They are not spiteful, or obdurate, or hopelessly prideful. They have been raised in a culture and an atmosphere that trains them to see the world in a useful but ultimately false way. The picture of the universe that was handed down to them is a flat one, containing only material truths. And since they rightly see that science *delivers* material truths, they conclude that science is true and the Bible is not. When the Soviet cosmonaut Gherman Titov visited the Seattle World's Fair in 1962, he reported that he had seen "no God or angels" on his journey to the stars. Posters celebrating this supposedly conclusive revelation went up across Russia, showing Titov grinning out into the empty reaches of space above a cheery caption in block letters: *Boga Nyet!* "There is no God."[3]

Titov's pronouncement betrays a common assumption: either the divine visage can be seen and touched with eyes and hands of flesh, somewhere in the neighborhood beyond earth's atmosphere, or it must not exist at all. Decades after the fall of the Soviet Union, these two narrow alternatives hem in men and women around the world before they ever even realize there are other options. Either something is physically true or it is absolutely false: this is the unseen assumption that clouds the vision of modernity's children. It will not be dispelled by simply insisting that scientific fact is not fact. People can see that it is. What they cannot see is how *incomplete* it is, how the discoveries of science are meaningless and ultimately incoherent unless informed by spiritual realities that can only be conveyed through the kinds of stories we find in Genesis.

"They say, / The solid earth whereon we tread / In tracts of fluent heat began, / And grew to seeming-random forms, / The seeming prey of cyclic storms, / Till at the last arose the man": even in the days of Alfred, Lord Tennyson, this sorrowful and barren picture of the world was taking hold of minds in pain. Today its power has only grown.[4] Honest, intelligent people of enormous talent—people like Einstein—abandon faith because they cannot see what it adds to their purely material picture of the world. And so it is that soul after soul is left bereft and directionless in a society that teaches them to treat their bodies like meat and their minds like a chemistry set. The hungry sheep look up, and are not fed.

Under these conditions, what believer's heart would not heave with pity? It would be a poor evangelist indeed who would simply leave a starving generation with no word of insight that can feed their hearts. If many turn away from God out of pride—or any of the other seven deadly sins—many others turn away from God because they see in Him no viable alternative to materialism.

At the dawn of the scientific age, Bishop George Berkeley fretted that readers would be led by the new philosophy of physics to come to the Bible looking for a purely material description of the world. "That a corporeal substance, which hath an absolute existence without the minds of spirits, should be produced out of nothing by the mere will of a spirit," was so implausible that it had "furnished the *atheists* and *infidels* of all ages, with the most plausible argument against a creation."[5] In this, he has been proven right: all over the world today, men and women are swindled out of the truths that pulse in their own hearts, taught to theorize their own souls out of existence, because they cannot prove them by an equation or point to them on a chart. If you want to reach or heal them, you won't do it by sniffing at the real discoveries of science. You'll do it by showing them that even those discoveries point beyond what can be measured or graphed—that science can't explain itself.

The hour is too late, and the issue too urgent, for people of good will to abandon natural theology or write off all science as a spiritual lost cause. At this stage in our history, science and theology desperately need each other. If they remain suspicious of one another, science will become an increasingly irrational and antihuman enterprise, sacrificing humanity itself on the altar of a misbegotten efficiency. Theology will find its protestations falling ever more softly on the ears of a generation made deaf to the things of the spirit.

But God and knowledge are not natural enemies: just the opposite. God's majesty remains in evidence all over the spectacular home he has built for his creatures. What I have been arguing in these pages is that we were not left here as orphans, crying out in anguish for an answer that will never come. The universe itself *is* the answer to the fact of our existence, and the light that streams from our minds meets a light that was infused in the world from its very foundation.

Once realize this and the physical world comes surging back to life at our touch. What we will find, if we look, is that the human soul has been invested with unfathomable honor, charged with gathering the world and its wheeling seasons, the cosmos and its dazzling architecture, into one unending hymn of praise. The world will meet our wondering minds with revelation after revelation, streaming forth from an inexhaustible source in the mind that made the world, and made us. And we shall see that it is good.

ACKNOWLEDGMENTS

Every book requires heroic patience from the people who love its author. I am exceptionally lucky in the grace that my family grants me, the love they give to sustain me. Joshua Herr, Ellen Klavan, Andrew Klavan, Faith Moore: thank you.

Harry Crocker and the entire Regnery team believed in me from day one; I could not ask for a better editor or for more unflagging support from these consummate professionals. I'm grateful as well to the team at Skyhorse, especially Elizabeth Kantor and Justin Fetterman, for steering me ably through a period of transition. In that process, Cynthia Angulo did a beautiful and meticulous job crafting the cover.

This book also needed a lot of sympathetic but critical readers; I had those too. First and foremost among them were my mother Ellen and my brother in all but birth, Jonathan Hay. Thanks also be to God for Joe Alto, Geoff Tanner, Jess Holmes, Emmet Penney, Michael Martin, Glen Williamson, and Michael Rezsler. Stephen C. Meyer and Emily Sandico of the Discovery Institute were instrumental—not just because of Steve's elegant and thoughtful preface, for which I'm deeply grateful, but also because of Emily's exceptionally perceptive notes and corrections. And thanks to Inez Stepman and Sam MacDonald for eleventh-hour help with sources on Signal and Twitter/X—proof that technology can be a blessing, not just a curse.

ENDNOTES

Introduction: From Dark to Dawn

1   See esp. David Bohm, *Wholeness and the Implicate Order* (London: Routledge, 1980), 248–52.

2   Richard Feynman, *The Feynman Lectures on Physics*, ed. Michael A. Gottlieb and Rudolph Pfeiffer (Pasadena: The California Institute of Technology, 1961), Vol. 1, Ch. 1. https://www.feynmanlectures.caltech.edu.

3   Richard Feynman, "The Value of Science," delivered at the National Academy of Science, 1955. https://calteches.library.caltech.edu/1575/1/Science.pdf.

4   Sarah Silverman, *We Are Miracles*. HBO, 2014. https://www.youtube.com /watch?v=xSiM5v-S_eY; "Sarah Silverman Wins for Writing for a Variety Special." Television Academy. https://www.youtube.com/watch?v=v246gNq IfTU. A distinct though related view is that we are all just lines or operations of a complex computer code: see for instance Lex Fridman (@lexfridman), "Humans are an API to ChatGPT. ChatGPT is an API to Python. Python is an API to C. C is an API to assembly. Assembly is an API to binary. Binary is an API to physics. Physics is an API to the machine that runs the universe. It's computation all the way down," Twitter/X, February 5, 2023. https://twitter .com/lexfridman/status/1622280318861418497?s=20&t=lkmEzipm1F72Y -Aol3xqFA.

5   "Revelation," *Midnight Mass* Season 1, episode 7, dir. Mike Flanagan. Netflix: September 24, 2021.

6   Rolf De Heer, *Bad Boy Bubby*, directed by Rolf De Heer (Roadshow Entertainment, 1993), 114 minutes.

7   Philip Pullman, *The Amber Spyglass* (Oxford: David Fickling Books, 2000), 31.

8   Carl Sagan, *The Cosmic Connection: An Extraterrestrial Perspective*, produced by Jerome Agel (New York: Dell Publishing Co., 1973), 148.

9   Neil deGrasse Tyson (@neiltyson), "Ever look up at night and feel small? Don't. Instead feel large. Atoms in our bodies trace to the remnants of exploded stars. We are Stardust. We are alive in the universe. And the universe is alive within us." Twitter/X, August 23, 2020. https://twitter.com/neiltyson /status/1297599280425861122?s=20&t=Q-ooNrtAfYbIgEeCyc3AGA.

10   u/Systemofalemon. "The skeleton isn't inside you, you're the brain so you're inside the skeleton." Reddit, 2016. https://www.reddit.com/r/Showerthoughts /comments/4dgpg9/the_skeleton_isnt_inside_you_youre_the_brain_so/. Previous expressions of the same thought had been less successful, and the "bone mech" concept was added on later. See further r/TeddyTehFrog. "Skeletons are not inside of us, we are inside a skeleton. We are a brain." 2015. https://www.reddit.com/r/Showerthoughts/comments/3evlhd/skeletons_are _not_inside_of_us_we_are_inside_a/; "Sometimes My Genius . . . It's Almost Frightening - you're piloting a bone mech that's using meat armor." Reddit, KnowYourMeme.https://knowyourmeme.com/photos/1731849-sometimes -my-genius-its-almost-frightening; Twitter Search, "bone mech meat armor," January 9, 2023. https://twitter.com/search?q=bone%20mech%20meat%20 armor&src=saved_search_click&f=top.

11   Karl Pilkington, *The Ricky Gervais Show*, Season 2, episode 4, "Onion." Aired February 4, 2011.

12   James (@kloogans), "pretty fucked up that my body, a chemistry lab made of meat, simply chooses to make me feel a little bit nervous for no reason." Twitter/X, October 5, 2022. https://twitter.com/kloogans/status/1577666614 908133378?s=20&t=Whh6DYN_s6zkKG8uJEHWSQ.

13   u/pasta_pants. "we're all just atoms and nothing we do matters in this universe." Reddit, r/ShowerThoughts, 2020. https://www.reddit.com/r/Shower thoughts/comments/aewxvz/were_all_just_atoms_and_nothing_we_do _matters_in/?utm_source=share&utm_medium=web2x&context=3; "We're all atoms. What's the point of living?" Quora, 2010. https://qr.ae/prsiCN; Bill Cravens, "What is the point in living if we are made of atoms and everything around us is an illusion?" Quora, 2018. https://qr.ae/prsiHN. See further u/ Cortxz. "what is the point of life? like what are we here to do? what is the objective? nothing ever really matters because we're all just atoms floating around in the universe. why are we here?" Reddit, r/3AMThoughts, 2020.

14 See e.g. Paul Bogowicz et al., "Trends and Variation in Antidepressant Prescribing in English Primary Care: A Retrospective Longitudinal Study" in *BJGP Open* 5, no. 4 (2021), doi: 10.3399/BJGPO.2021.0020; Julia Robinson, "Antidepressant Prescribing Up 6% in Last Three Months of 2020," *The PJ*, March 2021. https://pharmaceutical-journal.com/article/news/antidepressant -prescribing-up-6-since-2019.

15 Ruairi J. Mackenzie, "A Popular Theory About Depression Wasn't 'Debunked' by a New Review," *Technology Networks*, July 22, 2022. https://www.technology networks.com/neuroscience/articles/a-popular-theory-of-depression-wasnt -debunked-by-a-new-review-it-got-debunked-years-ago-363986; Pamela D. Pilkington et al., "The Australian Public's Beliefs about the Causes of Depression," *Journal of Affect Discord* 150, no.2 (2013): 356–62.

16 Mara Altman, "There Has Never Been a Better Time to be Short," *New York Times*, January 1, 2023. https://www.nytimes.com/2023/01/01/opinion/height -short.html; Catherine Clifford, "White House Is pushing Ahead Research to Cool Earth by Reflecting Back Sunlight," CNBC, October 13, 2022. https: //www.cnbc.com/2022/10/13/what-is-solar-geoengineering-sunlight-reflection -risks-and-benefits.html; Xiaomeng Han, "Can We Erase Painful Memories with Electroconvulsive Therapy?" *Harvard University Graduate School of Arts and Sciences Blog*, October 1, 2017. https://sitn.hms.harvard.edu/flash/2017 /can-erase-painful-memories-electroconvulsive-therapy/#:~:text=Figure%20 1%3A%20People%20with%20PTSD,memories%2C%20and%20therefore%20 help%20people,

17 Natasha Vita-More, *Transhumanist Manifesto, v4*, 2020. https://natashavita -more.com/transhumanist-manifesto/. See further Martine Rothblatt, *From Transgender to Transhuman* (Self-published, 2011); Nick Bostrom, "Why I Want to Be a Posthuman When I Grow Up," in *Medical Enhancement and Posthumanity*, ed. Bert Gordijn and Ruth Chadwick (New York: Springer, 2008), 107–37; n1x, "Gender Acceleration: A Blackpaper," October 31, 2018. https://vastabrupt.com/2018/10/31/gender-acceleration/.

18 Yuval Noah Harari, *Homo Deus (A Brief History)* (New York: Harper, 2015), 357.

19 See Aron Kheriaty, *The New Abnormal: The Rise of the Biomedical Security State* (Washington, DC: Regnery, 2022), 48–64.

Chapter 1

1 Homer, *Iliad* 14.201–302; see further Aristotle, *Metaphysics* 983b-84a; Peter J. Ahrensdorf, *Homer and the Tradition of Political Philosophy: Encounters with Plato, Machiavelli, and Nietzsche* (Cambridge, UK: Cambridge University Press, 2022), 39; Spencer Klavan, "The Philosopher and the Poet," *Law & Liberty* December 2022. https://lawliberty.org/book-review/the-philosopher -and-the-poet/.

2 *Enuma Elish* 1–5. On the date and context see Stephanie Dalley, *Myths from Mesopotamia: Creation, The Flood, Gilgamesh, and Others* (Oxford: Oxford University Press, 1989), 228–29; Paul-Alain Beaulieu, *A History of Babylon: 2200 BC–AD 75* (Chichester: Wiley-Blackwell, 2018), 161–62.

3 Aristotle, *Metaphysics* 983b.

4 Strictly speaking, these are "material monists": see Börje Bydén, "Some Remarks on the Text of Aristotle's Metaphysics," *Classical Quarterly* 55, no. 1 (2005): 108-9.

5 Plato, *Phaedo* 77a. Cf. Aristotle, *Physics* 244b.

6 Plato, *Phaedo* 65d.

7 Cf. e.g. Plato, *Meno* 100a, *Republic* 533e–4c, *Theaetetus* 184–86; Robert Loriaux, *Le Phédon de Platon* (Gembloux: Secrétariat des publications, Facultés universitaires, 1969), 167; Miles Burnyeat, "Plato on the Grammar of Perceiving," *Classical Quarterly* 26, no. 1 (1976): 45–46, 49–51.

8 Plato, *Symposium* 211d-e.

9 Aristotle, *Metaphysics* 987b.

10 Aristotle, *Metaphysics* 985b.

11 Parmenides, *On Nature* fr. 8.34, in Simplicius's *Commentary on Aristotle's Physics* 179.31. Cf. fr. 4 in Clemens of Alexandria, *Stromata* 6.2.23. See further James A. Weisheipl, "Comment," *The Concept of Matter in Modern Philosophy*, ed. Ernan McMullin (Notre Dame, IN: University of Notre Dame Press, 1963), 101–2.

12 For Aristotle's answer to the claim that no change really happens, see *Physics* 191b with C. J. Wolfe, "Plato's and Aristotle's Answers to the Parmenides Problem," *The Review of Metaphysics* 65, no. 4 (2012): 747–64, 754–58.

13 Aristotle, *Metaphysics* 984b.

14 The excitement of those days in Aristotle's Athenian school, the Lyceum, is captured in the story that he and his disciples were known for compulsively "walking about"—in Greek, *peripatein*—which is why they are called the "Peripatetics." See Diogenes Laertius, *Lives of the Eminent Philosophers* 5.2.

15   The conglomeration of these two Greek terms gives the idea its name: "hyle-morphism." See Aristotle, *Categories* 1–5; *Physics* 189b-91a; *Metaphysics* 1028a. Cf. Spencer Klavan, *How to Save the West* (Washington, DC: Regnery, 2023), 60.

16   William James, *The Principles of Psychology* (New York: Henry Holt & Co., 1918), Vol.1 Ch. 13, p. 488.

17   Aristotle, *De Anima* 432a.

18   In the technical language of Aristotle's "four causes," this means that the formal and final cause are one and the same for natural bodies: Aristotle, *Physics* 198a. Cf. *Metaphysics* 1013a-b.

19   This is why Thomas Kuhn writes that "animism has been transmuted" by the Aristotelian laws of motion. Thomas S. Kuhn, *The Copernican Revolution: Planetary Astronomy in the Development of Western Thought* (Cambridge, MA: Harvard University Press, 1957), 97. It is important that we cannot escape this kind of thinking—even though we try to talk about laws of motion as inanimate forces or tendencies, high school physics and chemistry teachers still talk as if certain atoms "want to" bond with others or "try to" achieve various electron configurations. See further pp. 122–24.

20   Aristotle, *Physics* 191b.

21   Aristotle *On the Heavens* 286b.

22   Tertullian, *Apology* 1. See further Tom Holland, *Dominion: How the Christian Revolution Remade the World* (New York: Basic Books, 2019), 131–33.

23   Isaiah 40:8.

24   See, for instance, Thomas S. Kuhn, *The Copernican Revolution: Planetary Astronomy in the Development of Western Thought* (Cambridge, MA: Harvard University Press, 1957), 107–8, on Augustine, *Enchiridion* 9.180–81; Lactantius, *Divine Institutions* Book III.

25   Aratus, *Phaenomena* 1–5. See Robin Hard, *Eratosthenes and Hyginus: Constellation Myths, with Aratus's Phaenomena* (Oxford: Oxford University Press, 2015), xx–xxi.

26   Aratus, *Phaenomena* 5 in Acts 17:28.

27   On the translation of Aristotle's scientific writings into Latin during the 1100s and 1200s, see John Losee, *Historical Introduction to the Philosophy of Science* (Oxford: Oxford University Press, 1972), 27–28.

28   Thomas Aquinas, *Of God and His Creatures* III.34. On the super- and sublunary realms see further Maimonides, *Guide of the Perplexed* 2.24.

29   See Aristotle, *Physics* 258b-9a, *Metaphysics* 1072a-b. Cf. *On the Heavens* 270b.

30 Thomas Aquinas, *Summa Theologica* Part 1 Question 2, article 3. This "argument from motion" is the first of Aquinas's five proofs that God exists—he calls it "the first and more apparent" of the ways to know that there is a God.

31 Dante, *Paradiso* 1.103–120, 33.145. See *The Divine Comedy of Dante Alighieri*, trans. Robert M. Durling, ed., Notes by Ronald L. Martinez and Robert M. Durling (Oxford: Oxford University Press, 1996–2011).

32 See Aristotle, *Physics* 208b, 215a, 266a–7b; *On the Heavens* 296b.

33 Jean Buridan, *Questions on the Eight Books of the Physics of Aristotle*, Book VIII, question 12, 1–3. The first of these arguments, the argument from contradiction, or *modus tollens*, had recently been formalized by Oxford's Robert Grosseteste. In general, Buridan is proceeding along a line of reasoning that would be crucial for the development of modern science. Gathering together multiple instances of the same occurrence—in this case, motion which continues after the inciting force has stopped—the philosopher tries to attempt a possible cause among them. The Franciscan friar John Duns Scotus outlined this "method of agreement." See John Losee, *A Historical Introduction to the Philosophy of Science* (Oxford: Oxford University Press, 1972), 29–34.

34 Jean Buridan, *Questions on the Eight Books of the Physics of Aristotle*, Book VIII, question 12, 4. An earlier form of this proposal can be found in John Philoponus, *Commentary on Aristotle's Physics* 214b (639–42), and a modified version was widespread among the medieval Arab philosophers: Shlomo Pines, "Les précurseurs musulmans de la théorie de l'impetus," *Archeion* 21, no. 3 (1938): 299–306.

35 See Plato, *Timaeus* 41c. Cf. Dante, *The Banquet* 69, 79–80.

36 Jean Buridan, *Questions on the Four Books of Aristotle's* On the Heavens and the Earth, Book II, Question 12.

37 Richard C. Dales, "The De-Animation of the Heavens in the Middle Ages," *Journal of the History of Ideas* 41, no. 4 (1980): 531–50. See especially 547–49.

38 Herbert Butterfield, *The Origins of Modern Science: Revised Edition* (New York: The Free Press, 1957), 19–20.

39 Nicole Oresme, *Le Livre du Ciel et du Monde*, II.3.300.

40 William of Ockham, *Summulae in Philosophae Naturalis* 3.5–7.

41 John Milton, "Sonnet 19: When I consider how my light is spent," lines 12–13.

Chapter 2

1   See also John John of Sacrbosco, *On the Sphere*, quoted in Edward Grant, *A Source Book in Medieval Science* (Cambridge, MA: Harvard University Press, 1974), 465.

2   See Alexander Jones, *A Portable Cosmos: Revealing the Antikythera Mechanism, Scientific Wonder of the Ancient World* (New York: Oxford University Press, 2017), 1–14. See Images 5 and 6.

3   For an ingenious argument that the mechanism could in principle have displayed all five planets known to antiquity (Mercury, Venus, Mars, Jupiter, and Saturn) plus the moon, the sun, and the stars, see Tony Freeth and Alexander Jones, "The Cosmos in the Antikythera Mechanism," ISAW Papers 4 (2012): http://doi.org/2333.1/xgxd26r7. A highly fictionalized version of mechanism features as the McGuffin in *Indiana Jones and the Dial of Destiny* (2023).

4   Cicero, *De Natura Deorum* 2.88. The speaker is Quintus Lucilius Balbus, who serves as Cicero's representative of Stoic teaching. Tony Freeth, "The Antikythera Mechanism 2. Is it Posidonius' Orrery?," *Mediterranean Archaeology and Archaeometry* 2, no. 2 (2002): 45–58 argues that the Antikythera mechanism may in fact be the remains of Posidonius's device. See further Alexander Jones, *A Portable Cosmos: Revealing the Antikythera Mechanism, Scientific Wonder of the Ancient World* (New York: Oxford University Press, 2017), 94. It is sometimes speculated that Geminus, Posidonius's fellow Rhodesian, may also have been the mechanism's creator. See Alex Nice, "Review of Alexander Jones, *A Portable Cosmos*," *Bryn Mawr Classical Review* (2018). https://bmcr.brynmawr.edu/2018/2018.05.16/.

5   See Thomas S. Kuhn, *The Copernican Revolution: Planetary Astronomy in the Development of Western Thought* (Cambridge, MA: Harvard University Press, 1957), 13–41.

6   See Paul Kriwaczek, *Babylon: Mesopotamia and the Birth of Civilization* (New York: Thomas Dunne Books, 2010), 197–98. Cf. Marc Van De Mieroop, *A History of the Ancient Near East, ca. 3000–323 BC*, 3rd ed. (Chichester: Wiley-Blackwell, 2016), 281–82 on the Enuma Anu Enlil.

7   Simplicius on Aristotle's *De Caelo* 2.43, 46.

8   Plato, *The Republic* 529a-e.

9   This is called "homocentric" motion, since all planets including the sun and moon are supposed to follow circular motions with the earth at their center. Thomas S. Kuhn, *The Copernican Revolution: Planetary Astronomy in the Development of Western Thought* (Cambridge, MA: Harvard University Press, 1957), 55–59. See Image 1.

10   These secondary circles, called "epicycles," are oriented around a moving center, whose trajectory is called a "deferent." See Thomas S. Kuhn, *The Copernican Revolution: Planetary Astronomy in the Development of Western Thought* (Cambridge, MA: Harvard University Press, 1957), 59–64. See Images 2–4.

11   See Pierre Duhem, *To Save the Phenomena*, trans. Edmund Dolan 1969 (Chicago: University of Chicago Press, 1908), 9 on Theon, *Astronomia* 34.

12   See Edward Rosen, *Three Copernican Treatises* (Mineola, NY: Dover Publications, 1939), 34–53.

13   See Ptolemy, *Almagest* I.7; John Losee, *A Historical Introduction to the Philosophy of Science* (Oxford: Oxford University Press, 1972), 18–19.

14   Copernicus, *De Revolutionibus*, Letter to Pope Paul III.

15   Ibid. The Pythagoreans named are Philolaus and Ecphantus. Copernicus also mentions Heraclides of Pontus, though it is a matter of some debate whether he owed anything to the heliocentric model proposed in the third century BC by Aristarchus of Samos, and if so how much. See Edward Rosen, "Aristarchos of Samos and Copernicus," *The Bulletin of the American Society of Papyrologists* 15, no. 1/2 (1978): 85–93.

16   Marsilio Ficino, *De Sole* ch.2. in *Marsilii Ficini Florentini Opera* (Basel: Henric Petrina, 1567), I, 966.

17   Copernicus, *De Revolutionibus* I.10. See Thomas S. Kuhn, *The Copernican Revolution: Planetary Astronomy in the Development of Western Thought* (Cambridge, MA: Harvard University Press, 1957), 130–31.

18   See further Arthur Koestler, *The Sleepwalkers: A History of Man's Changing Vision of the Universe* (London: Arkana, 1959), 151–54.

19   See Thomas S. Kuhn, *The Copernican Revolution: Planetary Astronomy in the Development of Western Thought* (Cambridge, MA: Harvard University Press, 1957), 149–50, 166–67; Edward Rosen, *Three Copernican Treatises* (Mineola, NY: Dover Publications, 1939), 37–38.

20   Galileo, *Sidereus Nuncius* 5v. Translation after Albert Van Helden, *Sidereus Nuncius, English Translation* (Chicago: University of Chicago Press, 1989). https://www.reed.edu/math/wieting/mathematics537/SideriusNuncius.pdf.

21   De Zúñiga, *In Job Commentaria* 9:5 (the passage referred to is actually 9:6, but it is mistakenly numbered in the commentary).

22   Thomas Aquinas, *Summa Theologica* Part I, question 68, article 3; Part III, question 57, articles 1 and 4. Nicholas Oresme, *Le Livre du ciel et du monde*, in Marshall Claggett, *The Science of Mechanics in the Middle Ages* (Madison: University of Wisconsin Press, 1959), 602–3. See further Edward Grant,

"Late Medieval Thought, Copernicus, and the Scientific Revolution," *Journal of the History of Ideas* 23 no.2 (1962): 210–11.

23 Copernicus, *De Revolutionibus* I.8, Virgil, *Aeneid* III.72.

24 Galileo, *Letter to the Grand Duchess of Tuscany* (Nashville, TN: Fordham University, Internet Modern History Sourcebook, ed. Paul Haslall). https://joelvelasco.net/teaching/3330/galileo-letter_to_grand_duchess.pdf

25 These and other documents from the case are extensively reproduced in Giorgio De Santillana, *The Crime of Galileo* (Chicago: University of Chicago Press, 1955). See especially 125–44 for a demonstration that Galileo was not prohibited, at this point, from teaching heliocentrism as a theory—only from affirming it as fact.

26 Galileo Galilei, *The Assayer*. See *The Essential Galileo*, ed. and trans. Maurice A. Finocchiaro (Indianapolis, IN: Hackett, 2008), 179–89.

27 See De Santillana, *The Crime of Galileo* (Chicago: University of Chicago Press, 1955), 246–47.

28 See *Galileo Galilei: Dialogue Concerning the Two Chief World Systems*, trans. Stillman Drake, ed. Stephen Jay Gould (New York: The Modern Library, 2001). And see further Edward Grant, "Late Medieval Thought, Copernicus, and the Scientific Revolution," *Journal of the History of Ideas* 23 no.2 (1962): 197–220, 216–17.

29 Thomas S. Kuhn, *The Copernican Revolution: Planetary Astronomy in the Development of Western Thought* (Cambridge, MA: Harvard University Press, 1957), 226.

Chapter 3

1 See Thomas S. Kuhn, *The Copernican Revolution: Planetary Astronomy in the Development of Western Thought* (Cambridge, MA: Harvard University Press, 1957), 212–14.

2 See further Melissa Cain Travis, *Thinking God's Thoughts: Johannes Kepler and the Miracle of Cosmic Comprehensibility* (Moscow, Idaho: Roman Roads Press, 2022), *passim* and esp. 152–56. When combined with Newton's principles of gravitation, Kepler's third law yields the formula $T^2=a^3$, where T is the time required for the planet to complete one revolution (the period) and a is the farthest possible distance from the center to the perimeter of the ellipse (the semimajor axis).

3 Kepler, *Apologia Tychonis contra Nicolaum Raymarum Ursum* in Kepler *Opera Omnia*, ed. C. Frisch (Frankfurt am Main and Erlangen: Heyder & Zimmer,

1858–1870), vol. 1 p. 239. In this Kepler was rejecting the tactful compromise between heliocentrism and geocentrism proposed by his mentor Tycho Brahe, in whose system all planets besides Earth orbited the sun, which in turn orbited Earth along with the moon and the stars. See Pierre Duhem, *To Save the Phenomena*, trans. Edmund Dolan 1969 (Chicago: University of Chicago Press, 1908), 67, 96–97, 101–2. It was Kepler who identified Osiander as the author of the preface to *De Revolutionibus* and dismissed it as a hopeless attempt to cover over the real meaning of Copernicanism.

4   Quoted and translated in Edwin Arthur Burtt, *The Metaphysical Foundations of Modern Physical Science*, 2nd ed. (New York: Harcourt, Brace, 1932), 48.

5   Bacon, *De Augmentiis Scientiarum* III.5. See *The Works of Francis Bacon: Philosophical Works 1*, ed. James Spedding, Robert Leslie Ellis, and Douglas Denon Heath (Cambridge, UK: Cambridge University Press, 2013), vol. 1.

6   This is why Arthur C. Clarke was able to make his famous rule that "any sufficiently advanced technology is indistinguishable from magic." Arthur C. Clarke, *Profiles of the Future: An Inquiry into the Limits of the Possible* (New York: Holt, Rinehart & Winston, 1962, rev. 1984), 14–36.

7   See C. S. Lewis, *The Abolition of Man* (New York: HarperOne, 1944), 76–77.

8   Francis Bacon, *Novum Organum*, ed. Lisa Jardine and Michael Silvertorn (Cambridge, UK: Cambridge University Press, 2000), Preface and Aphorisms I.39–44.

9   See Barbara Shapiro, *Probability and Certainty in Seventeenth-Century England: A Study of the Relationships Between Natural Science, Religion, History, Law, and Literature* (Princeton: Princeton University Press, 1983), 15–73, esp. 40. Cf. John Losee, *A Historical Introduction to the Philosophy of Science* (Oxford: Oxford University Press, 1972), 69–71.

10   See *The Alchemy Reader: From Hermes Trismegistus to Isaac Newton*, ed. Stanton J. Linden (Cambridge, UK: Cambridge University Press, 2003), 63–64.

11   A traditional view, notably expounded by Thomas Aquinas, held that prime matter was a logical abstraction rather than a physical reality. Thomas Aquinas, *Summa Theologica* Part I q. 66. See further Joseph Owens, "Matter and Predication in Aristotle," in *The Concept of Matter in Greek and Medieval Philosophy*, ed. by Ernan McMullin (Notre Dame, IN: The University of Notre Dame Press, 1963), 79–93, at 82–4, 92–3; James Dominic Rooney, *Material Objects in Confucian and Aristotelian Metaphysics: The Inevitability of Hylomorphism* (London: Bloomsbury, 2022), 66–68; Spencer Klavan, "There Is No Raw Material," in *The American Mind*, March 9, 2023. https://american mind.org/features/soul-dysphoria/there-is-no-raw-material/.

12 Hermes Trismegistus, *Emerald Tablet* 1–6. See further e.g. Khalid Ibn Yazid, *Secreta Alchymiae* 23, in *The Alchemy Reader: From Hermes Trismegistus to Isaac Newton*, ed. Stanton J. Linden (Cambridge, UK: Cambridge University Press, 2003), 27–28, 73.

13 See Bruce Janacek, *Alchemical Belief: Occultism in the Religious Culture of Early Modern England* (Pennsylvania: University of Pennsylvania Press, 2011), 37–38 on Thomas Tymme's *Dialogue Philosophicall*, 69–70.

14 Abraham Cowley, "To the Royal Society," 5. 93–98.

15 On the interest of early modern scientific revolutionaries in alchemy, see further Bernard Jaffe, *Crucibles: The Story of Chemistry from Ancient Alchemy to Nuclear Fission* (New York: Dover, 1930, 4th ed. 1976), 7.

16 See Barbara J. Shapiro, *Probability and Certainty in Seventeenth-Century England: A Study of the Relationships Between Natural Science, Religion, History, Law, and Literature* (Princeton: Princeton University Press 1983), 27–37.

17 Bacon, *Temporis Partus Masculus* in Benjamin Farrington, *The Philosophy of Francis Bacon* (Liverpool: Liverpool University Press 1962), 72. Cf. David Colclough, "Ethics and Politics in *The New Atlantis*" in *Francis Bacon's New Atlantis: New Interdisciplinary Essays*, ed. Bronwen Price (Manchester: Manchester University Press, 2002), 60–81, 67–69. On seventeenth-century debates about the efficacy of experiment for proof see further Steven Shapin and Simon Schaffer, *Leviathan and the Air-Pump: Hobbes, Boyle, and the Experimental Life* (Princeton: Princeton University Press, 1985), 125–43. Empirical data was famously defended as a criterion of truth in the twentieth century by Karl Popper, though always with the acknowledgement that every observation comes pre-loaded with interpretational assumptions. Karl Popper, *The Logic of Scientific Discovery*. (London: Routledge, 1935, English ed. 1959), 23–24, 89–90. Famously, Thomas Kuhn would argue that experimental additions to knowledge within a preexisting system are entirely and radically distinct from philosophical revolutions in how *all* experimental results are to be interpreted. Thomas Kuhn, *The Structure of Scientific Revolutions* (Chicago: University of Chicago Press, 1962, expanded edition 1970), 103–8.

18 Descartes, *Discourse on Method* Part V.

19 Democritus, *Fragments* 9 and 49, in Galen, *On the Elements according to Hippocrates* I.2 and Sextus Empiricus, *Adversus Mathematicos* 7.135. Cf. *Hypotheses of Pyrrhonism* I.213–14; Diogenes Laertius, *Lives of Eminent*

Philosophers 9.34–45; Leucippus, *Fragment* 1 in Diogenes Laertius, *Lives of Eminent Philosophers* 9.30–33; *testimonium* 8, Theophrastus, *Opinions of the Natural Philosophers* I, fragment 8.19–21 (*apud* Simplicius on Aristotle's *Physics* 28.4). For Epicurus see e.g. Diogenes Laertius, *Lives of Eminent Philosophers* 10.45; 10.88–90, and see further James A. Weisheipl, "Comment," in *The Concept of Matter in Modern Philosophy*, ed. Ernan McMullin (Notre Dame, IN: University of Notre Dame Press, 1963), 100–103; Lucretius, *De Rerum Natura* 2.1023–1174.

20 See further Aristotle, *On Generation and Corruption* I.8 (325a–6a); Edward Hussey, "On Generation and Corruption I.8," in *Aristotle's On Generation and Corruption I*, ed. Frans de Haas and Jaap Mansfeld (Oxford: Oxford University Press, 2004), 258–61; Marie Boas Hall, "Matter in Seventeenth-Century Science," in *The Concept of Matter in Modern Philosophy*, edited by Ernan McMullin (Notre Dame, IN: University of Notre Dame Press, 1963), 77–8.

21 Descartes, *Principia Philosophiae* II.4. See further *The World* Chapter 6.

22 Descartes, *Discourse on Method* Part IV, *Principia Philosophiae* I.13–27.

23 See Richard J. Blackwell, "Descartes' Concept of Matter" in *The Concept of Matter in Modern Philosophy, ed. by Ernan McMullin* (Notre Dame, IN: University of Notre Dame Press, 1963), 59–75, 60–67.

24 Descartes, *The World* Chapter 5.

25 See e.g. Meyrick H. Carré, "Pierrre Gassendi and the New Philosophy," in *Philosophy* 33, no. 125 (April, 1958): 115; B. J. T. Dobbs, "Newton's Alchemy and His Theory of Matter," in *Isis* 73, no. 4 (December, 1982): 513; Thomas M. Lennon, *The Battle of the Gods and Giants: The Legacies of Descartes and Gassendi, 1655–1715* (Princeton: Princeton University Press, 1993), 7, 138–40.

26 On Descartes's objections to atomism as a metaphysical proposition—that there should be any truly indivisible body—see Thomas M. Lennon, *The Battle of the Gods and Giants: The Legacies of Descartes and Gassendi, 1655–1715* (Princeton: Princeton University Press, 1993), 202–4. On the ubiquity of particulate and corpuscular thought during this period, including in Descartes, see John Losee, *A Historical Introduction to the Philosophy of Science* (Oxford: Oxford University Press, 1972), 24–25.

27 In modern terms, this second law is expressed by the simple equation $F=ma$, where F and a are vectors describing, respectively, the force applied and the consequent acceleration of a body with mass m.

28 Newton, "Hypothesis Explaining the Properties of Light" and *Opticks* Query 31(23). See *Isaac Newton's Papers and Letters on Natural Philosophy*, ed. I.

Bernard Cohen, (Cambridge, UK: Cambridge University Press, 1958), 185; Ernan McMullin, *Newton on Matter and Activity* (Notre Dame, IN: University of Notre Dame Press, 1978), 75–109.

29 Newton, *Principia Mathematica* Definition IV.

30 Newton, *General Scholium* to the 1713 edition of the *Principia Mathematica*.

31 See Stephen C. Meyer, *Return of the God Hypothesis* (New York: HarperOne, 2021), 621.

32 Newton, *Principia Mathematica* (1687), Hypothesis III. See B. J. T. Dobbs, "Newton's Alchemy and His Theory of Matter," in *Isis* 73, no. 4 (December, 1982): 511–28, 514.

33 Newton, 1716 draft of projected addition to *Opticks* III. See further *Opticks* Query 31 and Ernan McMullin, *Newton on Matter and Activity* (Notre Dame, IN: University of Notre Dame Press, 1978), 143 n. 128.

34 Newton, *Principia Mathematica*, General Scholium (1687).

35 See Thomas S. Kuhn, *The Copernican Revolution: Planetary Astronomy in the Development of Western Thought* (Cambridge, MA: Harvard University Press, 1957), 215, 244–45, 254. Another possibility at the time was that magnetic force could account for planetary motion—William Gilbert had argued in *De Magnete* (1600) that the earth itself was one giant lodestone.

36 Newton's commentary is translated in B. J. T. Dobbs, *The Janus Faces of Genius: The Role of Alchemy in Newton's Thought* (Cambridge, UK: Cambridge University Press, 1991), 276–77. See further *The Alchemy Reader: From Hermes Trismegistus to Isaac Newton*, ed. Stanton J. Linden (Cambridge, UK: Cambridge University Press, 2003), 243–47.

37 Bacon, *Of the Wisdom of the Ancients* XVII: "Cupid, or The Atom," in *The Works of Francis Bacon: Literary and Professional Works*, ed. James Spedding, Robert Leslie Ellis, and Douglas Denon Heath (Cambridge, UK: Cambridge University Press, 2013), vol. 6.

Chapter 4

1 Pierre-Simon Laplace, *Essai philosophique sur les probabilités* (Paris: Bachelier, 1840), 4.

2 The idea in its outline was not unique to Laplace; it was a commonplace among the enlightenment philosophes of France, such as Condorcet and Diderot. See Marij van Strien, "On the Origins and Foundations of Laplacian Determinism," in *Studies in History and Philosophy of Science* 45, no. 1 (2014): 24–31, 26–30. This philosophical determinism found mathematical support

in the work of men like Pierre Louis Maupertuis, Joseph-Louis Lagrange, and Leonhard Euler, who among them developed what is now called the "stationary-action principle" or "principle of least action," relating the partial derivative of a system's Lagrangian (kinetic minus potential energy) with respect to the position of a particle within it, to the time integral of the partial derivative of the Lagrangian with respect to that particle's velocity. In other words: . This relatively simple calculus for charting and predicting the movements of particles over time is one of the early triumphs of classical physics.

See Leonard Susskind and George Hrabovsky, *The Theoretical Minimum: What You Need to Know to Start Doing Physics* (New York: Basic Books, 2013), 105–27.

3   Pierre-Simon Laplace, "Recherches sur l'intégration des différentielles aux différences finies et sur leur application à l'analyse des hasards," *Mémoires de mathématiques et de physiques présentés à l'Académie Royale des Sciences par divers savants*, vol. 7 (Paris: Imprimerie Royale, 1773), 37–233. Available online at https://www.biodiversitylibrary.org/item/86762#page/7/mode/1up.

4   See further Frederick Gregory's lecture "Consolidating Newton's Achievement" in *The History of Science: 1700–1900* (Chantilly, VA: The Teaching Company, 2013), 10:30–14:10. It has become commonplace to suggest that Laplace's discovery replaced Newton's own idea that God must regularly intervene to adjust or interrupt the laws of nature. But Stephen C. Meyer has shown convincingly that Newton only invoked God to explain the *existence* of the universe and its laws, not to make up deficiencies of the kind Laplace was solving: see *Return of the God Hypothesis* (New York: HarperOne, 2021), 642–48.

5   Lucretius, *De Rerum Natura* 2.1023–1174. See further Diogenes of Oinoanda, *fragment* 63.

6   See *The Hippocratic Regimen* chs. 25 and 35; cf. Theophrastus *De Sensu* chs. 10–11; Empedocles *fragment* 105; Jacques Jouanna, "The Theory of Sensation, Thought, and the Soul in the Hippocratic Treatise Regimen: Its Connections with Empedocles and Plato's *Timaeus*," in *Greek Medicine from Hippocrates to Galen: Selected Papers* (Leiden: Brill, 2012), 195–227, 208–18.

7   Hobbes, *Leviathan, Or: The Matter, Form, and Power of a Common-Wealth Ecclesiastical and Civil* (London: Andrew Cooke, 1651), introduction. See selections from *Leviathan* by Richard S. Peters, ed. by Michael Oakeshott, (Collier Macmillan, 1962), 19.

8   Descartes, *L'Homme et la formation du foetus* (Paris: Girard, 1677), 1–2, 97–98.

9   Lavoisier made this discovery with the help of the English schoolmaster
    Joseph Priestley.

10 The discovery was presented to the Academy in 1784 as the *Mémoire sur la
    chaleur*. See further Henry Guerlac, "Chemistry as a Branch of Physics:
    Laplace's Collaboration with Lavoisier," in *Historical Studies in the Physical
    Sciences* 7 (1976): 193–276.

11 See *Oeuvres de Lavoisier, publiées par les soins du Ministre de l'Instruction
    Publique*, vol. 2 (Paris: Imprimerie Imperiale, 1862), 546–56.

12 Voltaire, *Lettres Philosophique*, "Chancellor Bacon" (1733) in *The Works of
    Voltaire*, vol. 19 (Philosophical Letters) (Paris: E.R. DuMont, 1901).

13 Alexander Pope, "Epitaph on Sir Isaac Newton" 1–2.

14 See Julia L. Epstein, "Voltaire's Myth of Newton" *Pacific Coast Philology* 14
    (1979): 27–33.

15 *London Times*, September 10, 1792.

16 Wordsworth, *Prelude* 11.290–91.

17 See Sanja Perovic, *The Calendar in Revolutionary France: Perceptions of Time
    in Literature, Culture, Politics* (Cambridge, UK: Cambridge University Press,
    2012), 31, 87; cf. Rodney Stark, *Bearing False Witness: Debunking Centuries of
    Anti-Catholic History* (West Conshohocken, PA: Templeton Press, 2016), 198.

18 "Loi relative à la formation d'un Bureau des Longitudes, June 25th, 1795,"
    L'Observatoire de Paris: https://www.imcce.fr/institut/histoire-patrimoine
    /messidor.

19 Lavoisier was not so lucky.

20 Napoleon is said to have written this report in dismissing Laplace from his
    position as Minister of the Interior. See Étienne Ghys, "Napoléon Bonaparte
    et la science," Institut de France. May 4, 2021, https://youtu.be/zUlzqJaeCXo;
    Charles Coulston Gillispie, *Pierre-Simon Laplace: A Life in Exact Science*
    (Princeton: Princeton University Press, 1997), 176.

21 See Walter W. R. Ball, *A Short Account of the History of Mathematics* (New
    York: Macmillan 1893), 425–27; cf. "Napoleon Laplace Anecdote" at EoHT.
    info (http://archive.today/N3xFF).

22 Humphry Davy, *Researches Chemical and Philosophical Chiefly Concerning
    Nitrous Oxide, or Diphlogisticated Nitrous Air, and Its Respiration* (London: J.
    Johnson, 1800), 485–89.

23 Humphry Davy, *Researches Chemical and Philosophical Chiefly Concerning
    Nitrous Oxide, or Diphlogisticated Nitrous Air, and Its Respiration* (London: J.
    Johnson, 1800), 495–96, 516–17.

24 See John Dalton, *A New System of Chemical Philosophy Parts I and II* (Manchester: R. Bickerstaff, 1808–1810), *passim* but esp. plates at 218–19. Cf. Bernard Jaffe, *Crucibles: The Story of Chemistry from Ancient Alchemy to Nuclear Fission* (New York: Dover, 1930, 4th ed. 1976), 89–95. The wooden spheres that Dalton used for demonstration can be viewed at the website of the Science Museum Group, Object Numbers Y.1997.6.53 (https://collection.sciencemuseumgroup.org.uk/objects/co8410972/five-wooden-molecular-model-balls-molecular-model) and 1949–21 (https://collection.sciencemuseumgroup.org.uk/objects/co12617/wooden-atomic-models-used-by-john-dalton-1800-1844-ball-and-spoke-models-atomic-models).

25 Amadeo Avogadro, "Essai d'une manière de déterminer les masses relatives des molécules élémentaires des corps, et les proportions selon lesquelles elles entrent dans ces combinaisons," in *Journal de Physique* 73 (1811): 58–76.

26 See further Stephen Gaukroger, *Civilization and the Culture of Science: Science and the Shaping of Modernity 1795–1935* (vol. 4) (Oxford: Oxford University Press, 2020), 259–66, 356–64.

27 Charles Darwin, *The Origin of Species and The Voyage of the Beagle* (New York: Everyman's Library, 1859, reissued 2003), 601–2.

28 Gregor Mendel, "Versuche über Pflanzen-Hybriden," *Verhandlungen des naturforschenden Vereines in Brün*, Bund IV (1865): 3–47, 34–35. See further Ute Deichmann, "Gemmules and Elements: On Darwin's and Mendel's Concepts and Methods in Heredity," *Zeitschrift Für allgemeine Wissenschaftstheorie* 41, no. 1 (2010): 85–112, 99–100.

29 See Andrew Scull, *Desperate Remedies: Psychiatry's Turbulent Quest to Cure Mental Illness* (Cambridge, MA: The Belknap Press, 2022), 15–17.

30 Ivan Sechenov, *Reflexes of the Brain*, trans. S. Belskii (Cambridge, MA: MIT Press, 1965), 3–33. See further Alexey Vdovin, "Dostoevsky, Sechenov, and the Reflexes of the Brain: Towards a Stylistic Genealogy of Notes from Underground," in *Dostoevsky at 200: The Novel in Modernity* (Toronto: University of Toronto Press, 2021), 99–117, 104; Gary Saul Morson, *Wonder Confronts Certainty: Russian Writers on the Timeless Questions and Why Their Answers Matter* (Cambridge, MA: Harvard University Press, 2023), 25.

31 Léon Dumont, "De l'habitude," *Révue Philosophique*, Tome I (1876): 321–66, 324.

32 See William James, *Principles of Psychology* vol. 1 (New York: Henry Holt, 1890), 105–6.

33 Sigmund Freud, "Instincts and their Vicissitudes" (1915) in *The Stanford Edition of the Complete Psychological Works of Sigmund Freud*, Vol. 14, trans. and ed. James Strachey (London: The Hogarth Press, 1957), 125.

34 Sigmund Freud, "The Question of a Weltanschauung" (1933), in *The Freud Reader*, ed. Peter Gay (New York: W.W. Norton, 1989), 784 (cf. 789).

Chapter 5

1 Phil Springer (music) and Carolyn Leigh (lyrics), "(How Little it Matters) How Little We Know." 1956. Released on *Sinatra's Sinatra*, Capitol Records, 1963. Vinyl.

2 Marvin Fisher (music) and Jack Segal (lyrics), "Something Happens to Me," *Night of the Quarter Moon*, Capitol Records, 1958. Vinyl LP.

3 King Friedrich Wilhelm III, "Union Decree," Order-in-Council. Potsdam, September 27, 1817. Printed in the archives of the International Lutheran Council: https://ilconline.wpenginepowered.com/wp-content/uploads/2017/10/Union-Decree-Frederick-William-III.pdf (accessed November 28, 2023).

4 See further Thomas Nipperdey, *Germany from Napoleon to Bismarck: 1800–1866* (Princeton, NJ: Princeton University Press, 2014), 356.

5 Friedrich Nietzsche, *Daybreak* (1881), ed. Maudemarie Clark and Brian Leiter, trans. R. J. Hollingdale (Cambridge, UK: Cambridge University Press, 1997), 30 (§42), 32 (§49), 50 (§86), 70 (§114). Cf. 49 (§83), 77 (§122). See further *Ecce Homo* Ch. 3 and *Human, All Too Human* §1.

6 Friedrich Nietzsche, *Daybreak* (1881), ed. Maudemarie Clark and Brian Leiter, trans. R. J. Hollingdale (Cambridge, UK: Cambridge University Press, 1997), 14 (§14); Friedrich Nietzsche, *The Gay Science* (1882–1887), ed. Bernard Williams, trans. Josephine Nauckhoff (Cambridge, UK: Cambridge University Press, 2001), 119–20 (§125). Cf. 109 (§108) and the later Book V, 199 (§343).

7 Karl Marx and Friedrich Engels, *The German Ideology* (1845–1846), *The Marx-Engels Reader*, ed. Robert C. Tucker (New York: W.W. Norton, 1978), 149–50. On the composition see 66 and 146, and in defense of the accurate representation of Marx's views in the text see Thomas Sowell, *Marxism: Philosophy and Economics* (Rock Hill, SC: Quill, 1985), chapters 3–4.

8 See further Friedrich Engels, *Anti-Dühring* (1877), part III, chapter 2 (https://www.marxists.org/archive/marx/works/1877/anti-duhring/ch24.htm).

9 Karl Marx and Friedrich Engels, *The German Ideology* (1845–1846), in *The Marx-Engels Reader*, ed. Robert C. Tucker (New York: W.W. Norton, 1978), 154. See further Erich Fromm, *Marx's Concept of Man* (New York: Frederick Ungar, 1961), 1–85, esp. §2.

10   Karl Marx, *Kaptial* I.III.vii.1 (1867), in *The Marx-Engels Reader*, ed. Robert C. Tucker (New York: W. W. Norton, 1978), 345–46. See further 90–91.

11   Leon Trotsky, *Dictatorship vs. Democracy (Terrorism and Communism): A Reply to Karl Katusky* (New York: Workers Party of America, 1922), ch. 4: "Terrorism." See further Orlando Figes, *A People's Tragedy: The Russian Revolution 1891–1924* (New York: Viking, 1996), 641; Hugh Phillips, "The War Against Terrorism in Late Imperial and Early Soviet Russia," in *Enemies of Humanity: The Nineteenth-Century War on Terrorism*, ed. Isaac Land (New York: Palgrave Macmillan 2008), 219.

12   *Sovetskaya intelligentsiya* (Moscow: Izd-Vo Polit. Lit-Ry, 1977), 45 as cited in Richard Stites, *Revolutionary Dreams: Utopian Vision and Experimental Life in the Russian Revolution* (New York: Oxford University Press, 1989), 71. See further Gary Saul Morson, *Wonder Confronts Certainty: Russian Writers on the Timeless Questions and Why Their Answers Matter* (Cambridge, MA: Harvard University Press, 2023), 68–69.

13   See Victoria S. Frede, "Materialism and the Radical Intelligentsia: The 1860s" in *A History of Russian Philosophy, 1830–1930*, ed. G. M. Hamburg and Randal A. Poole (Cambridge, UK: Cambridge University Press, 2010), 78, cited in Gary Saul Morson, *Wonder Confronts Certainty: Russian Writers on the Timeless Questions and Why Their Answers Matter* (Cambridge, MA: Harvard University Press, 2023), 131.

14   See *Proletarskaya Kultura* 3 (August 1918), 1–18 and Richard Stites, *Revolutionary Dreams: Utopian Vision and Experimental Life in the Russian Revolution* (New York: Oxford University Press, 1989), 149–52.

15   *Proletarskaya Kultura* 9/10 (1919), 35–45.

16   Nikolai Krylenko, *Za Pyat Let* (1918–1922), 79, cited in Aleksandr Solzhenitsyn, *The Gulag Archipelago* vol. 1 (1973), trans. Thomas P. Whitney (New York: HarperCollins, 2007), 308–9. A gruesome catalog of Stalinist and pre-Stalinist violence under Soviet rule can be found in Gary Saul Morson, *Wonder Confronts Certainty: Russian Writers on the Timeless Questions and Why Their Answers Matter* (Cambridge, MA: Harvard University Press, 2023), 171–4.

17   Alan Turing, "On Computable Numbers, with an Application to the Entscheidungsproblem," *Proceedings of the London Mathematical Society* (1936), 230–32.

18   Alan Turing, "Intelligent Machinery" (National Physical Laboratory, 1948), 4. https://www.npl.co.uk/getattachment/about-us/History/Famous-faces/Alan-Turing/80916595-Intelligent-Machinery.pdf?lang=en-GB.

19 See further Gerard O'Regan, *A Brief History of Computing*, 3rd ed. (Switzerland: Springer, 2021), 33–35.

20 Alan Turing, "Intelligent Machinery" (National Physical Laboratory, 1948), 4. https://www.npl.co.uk/getattachment/about-us/History/Famous-faces/Alan-Turing/80916595-Intelligent-Machinery.pdf?lang=en-GB. See further Spencer Klavan, "The Exterior Darkness," *The American Mind*, June 19, 2023. https://americanmind.org/features/the-exterior-darkness/.

21 See Alan Turing, "Computing Machinery and Intelligence," in *Mind* 59, no. 236 (1950): 433–60, esp. §6(4).

22 See James Poulos, *Human Forever: The Digital Politics of Spiritual War* (Canonic, 2021), 18–19; David Bowie, Interview with Jeremy Paxman. *Newsnight*, BBC 2 (December 3, 1999). Archived at https://www.davidbowienews.com/2015/12/jeremy-paxman-2000/.

23 Elise Bohan, *Future Superhuman: Our Transhuman Lives in a Make-or-Break Century* (New South Wales: NewSouth Publishing, 2022), 11, 23–25. See further Mary Harrington, "Human Matters," in *The American Mind* (March 8, 2023): https://americanmind.org/features/soul-dysphoria/human-matters; *Feminism Against Progress* (Washington, D.C.: Regnery, 2023), 133–62.

24 Executive Office of the President. September 12, 2022. "Advancing Biotechnology and Biomanufacturing Innovation for a Sustainable, Safe, and Secure American Bioeconomy" (Executive Order 14081). https://www.federalregister.gov/documents/2022/09/15/2022-20167/advancing-biotechnology-and-biomanufacturing-innovation-for-a-sustainable-safe-and-secure-american.

25 Flo Crivello (@Altimor) and Elon Musk (@elonmusk), Twitter exchange on March 25–26, 2023. See https://twitter.com/elonmusk/status/1640183652800757766. Archived at https://archive.is/q1cAH.

26 See e.g. Dan Witters, "U.S. Depression Rates Reach New Highs," Gallup (May 17, 2023): https://news.gallup.com/poll/505745/depression-rates-reach-new-highs.aspx; Laura Dattner, "Youth Suicide and Attempted Suicide Reported to Poison Control Centers Increased During the COVID-19 Pandemic," *Pediatrics Nationwide* (April 04, 2023): https://pediatricsnationwide.org/2023/04/04/youth-suicide-and-attempted-suicide-increased-during-the-covid-19-pandemic/. Cf. Hans Jonas, *The Phenomenology of Life: Toward a Philosophical Biology* (Evanston, IL: Northwestern University Press, 1966), 108–34.

27 See nix, "Gender Acceleration: A Blackpaper," *The Anarchist Library* (2018). https://theanarchistlibrary.org/library/nix-gender-acceleration-a-blackpaper; cf. Douglas Murray, *The Madness of Crowds* (London: Routledge, 2019), 225, with video at https://archive.org/details/olson-kennedy-breasts-go-and-get -them. For a very mild summation of the appalling consequences of trans- gender surgery, acknowledged even by a sympathetic source, see "What Are the Risks of Transmasculine Bottom Surgery?" *The American Society of Plastic Surgeons* (2023): https://www.plasticsurgery.org/reconstructive-procedures /transmasculine-bottom-surgery/safety#:~:text=The%20possible%20risks %20of%20transmasculine,abnormal%20connections%20between%20the%20 urethra; "what are the risks of transfeminine bottom surgery?" https://www .plasticsurgery.org/reconstructive-procedures/transfeminine-bottom-surgery /safety#:~:text=The%20possible%20risks%20of%20transfeminine,the%20 skin%2C%20painful%20intercourse%20and. Accessed November 28, 2023. Archived at https://archive.is/UaFCE and https://archive.is/vnrPi.
28 Eliezer Yudkowsky, "Pausing AI Developments Isn't Enough. We Need to Shut It All Down," *The Economist* (March 29, 2023): https://time.com/6266923 /ai-eliezer-yudkowsky-open-letter-not-enough/.
29 Bertrand Russell, "The Study of Mathematics," *The New Quarterly* 1 (1907), reprinted in *Mysticism and Logic and Other Essays* (London: Longmans, Green, and Co., 1918), 60.

Chapter 6

1 Michael Faraday, "Thoughts on Ray-Vibrations," *The London, Edinburgh and Dublin Philosophical Magazine and Journal of Science* 28, no. 188 (1846): 345.
2 Ibid., 348.
3 Ibid., 348.
4 James Clerk Maxwell, "On Physical Lines of Force," *Philosophical Magazine* 23 (1862): 12–24, 85–95, 22. Emphasis original. See further James Clerk Maxwell, *Scientific Papers of James Clerk Maxwell*, vol. 1, ed. W. D. Niven (Cambridge, UK: Cambridge University Press, 1891), 500; Bruce J. Hunt, "Maxwell, Measurement, and the Modes of Electromagnetic Theory" *Historical Studies in the Natural Sciences* 45, no. 2 (2015), 314; Ann Breslin and Alex Montwill, *Let There Be Light: The Story of Light from Atoms to Galaxies* (London: Imperial College Press, 2013), 307–8. Maxwell's calculations are presented in full in James Clerk Maxwell, "A Dynamical Theory of the Electromagnetic Field" (presented 1864) in *Philosophical Transactions of the*

Royal Society of London 155 (1865): 459–512, 497–99. A more exact modern calculation of the speed of light is 299,792,458 meters per second.

5 See Ann Breslin and Alex Montwill, *Let There Be Light: The Story of Light from Atoms to Galaxies* (London: Imperial College Press, 2013), 308.

6 Albert Einstein, "Notes for an Autobiography," *Saturday Review of Literature* (November 26, 1949), 9–12.

7 Albert Einstein, "Zur Elektrodynamik bewegter Körper," *Annalen der Physik* 17 (1905): 891.

8 See Albert Einstein, *Relativity: The Special and General Theory, 100th Anniversary Edition*, with commentaries and background material by Hanoch Gutfreund and Jürgen Renn (Princeton: Princeton University Press, 2015), 21–22.

9 The mathematics for making these adjustments were already to hand—the Dutch physicist Hendrik Lorentz had developed them in an effort to describe how the ether might behave as a substance in space. Einstein uses the Lorentz transformations to account for the behavior of time and space in a frame of reference that is moving with uniform translational motion relative to another in "Zur Elektrodynamik bewegter Körper," *Annalen der Physik* 17 (1905): 891–921, 897–907.

10 See Albert Einstein, *Relativity: The Special and General Theory, 100th Anniversary Edition*, with commentaries and background material by Hanoch Gutfreund and Jürgen Renn (Princeton: Princeton University Press, 2015), 56–61.

11 Arthur S. Eddington, "The Internal Constitution of the Stars," *Nature* 106 (1920): 14–20.

12 Aristotle, *De Anima* 431a.

13 Alfred Korzybski, *Science and Sanity: An Introduction to Non-Aristotelian Systems and General Semantics* (New York: The Science Printing Company, 1933), 58.

14 David Tong, "Quantum Fields: The Real Building Blocks of the Universe," The Royal Institution (February 15, 2017): 10:50–11:06. https://www.youtube.com/watch?v=zNVQfWC_evg.

15 Albert Einstein, "Die Feldgleichungen der Gravitation," *Sitzungsberichte der Preussischen Akademie der Wissenschaften zu Berlin* (1915): 844–47. Einstein presented this as a physical explanation for the equivalence of inertial mass—a measure of a given body's resistance to force—and gravitational mass—a measure of that body's susceptibility to gravitational pull. He was able to add

time as a variable factor along with the three spatial coordinates thanks to the ingenious mathematics of Hermann Minkowski. For an intuitive depiction of how this works see Albert Einstein, *Relativity: The Special and General Theory, 100th Anniversary Edition*, with commentaries and background material by Hanoch Gutfreund and Jürgen Renn (Princeton: Princeton University Press, 2015), 80–83.

16 Albert Einstein, "Notes for an Autobiography," *Saturday Review of Literature* (November 26, 1949): 11.

17 Albert Einstein, *Relativity: The Special and General Theory, 100th Anniversary Edition*, with commentaries and background material by Hanoch Gutfreund and Jürgen Renn (Princeton: Princeton University Press, 2015), 80–83.

18 "The 1919 Eclipse Results that Verified General Relativity and their Later Detractors: A Story Re-Told," in *Notes and Records: The Royal Society Journal of the History of Science* 76, no. 1 (2022): https://doi.org/10.1098/rsnr .2020.0040.

19 See Richard Rhodes, *The Making of the Atom Bomb* (New York: Simon & Schuster, 1986), 40.

20 See J.G. Crowther, *The Cavendish Laboratory 1874–1974* (New York: Science History Publications, 1974), 123.

21 See G. K. T. Conn and H. D. Turner, *The Evolution of the Nuclear Atom* (London: Iliffe Books, 1965), 136–37.

22 The "solar system" model of the atom was developed independently by Hantaro Nagaoka and Ernest Rutherford: see See Richard Rhodes, *The Making of the Atom Bomb* (New York: Simon & Schuster, 1986), 50–51.

23 Maxwell posed this problem as a thought experiment that became known as "Maxwell's demon," an all-knowing entity to rival the one imagined by Laplace. But whereas Laplace's demon simply observed the predictable motion of particles, Maxwell's challenged it by transferring one high-energy particle of gas from an area with lower average temperature to another with a higher average. The result would violate the second law, but it was physically quite possible. See James Clerk Maxwell, *Theory of Heat* (London: Longmans Green, 1871), 328–29 in the 9th English edition.

24 See Ludwig Boltzmann, *Vorlesungen über Gastheorie* (Leipzig: Barth, 1896), 38–47. Cf. *Lectures on Gas Theory*, trans. S. G. Brush (Berkeley: University of California Press, 1964), 55–62.

25 See Philip Ball, *Beyond Weird: Why Everything You Thought You Knew about Quantum Physics Is Different* (London: Vintage, 2018), 27–28.

26 Max Planck, Letter to Lorentz, 7 October 1908. Quoted in Thomas S. Kuhn, *Black-Body Theory and the Quantum Discontinuity, 1894–1912* (Chicago: University of Chicago Press, 1978), 197–98.

27 See Max Planck, "Zur Theorie des Gesetzes der Energieverteilung im Normalspectrum," in *Verhandlungen der Deutschen Physikalischen Gesellschaft* 2 (1900): 237. Planck's and his contemporaries' often reluctant journey to accepting the constant is detailed in Thomas S. Kuhn, *Black-Body Theory and the Quantum Discontinuity, 1894–1912* (Chicago: University of Chicago Press, 1978), 188–205.

28 Max Planck, discussion following Einstein's "Über die Entwicklung unserer Anschauungen über das Wesen und die Konstitution der Strahlung," *Physikalische Zeitschrift* 10 (1909): 825.

29 Albert Einstein, "Über einen die Erzeugung und Verwandlung des Lichtes betreffenden heuristischen Gesichtspunkt," *Annalen der Physik* 17 (1905), 132–48.

30 See further Max Planck, "Zur Theorie des Gesetzes der Energieverteilung im Normalspectrum," *Verhandlungen der Deutschen Physikalischen Gesellschaft* 2 (1900): 237.

31 Louis de Broglie, "Ondes et Quanta," Institut de France, Académie des sciences Comptes Rendu 177 (1923): 507–10. See further Louis de Broglie, *Recherches sur la Théorie des Quanta* (Faculty of Sciences: Paris University PhD Thesis, 1924).

32 Gary Saul Morson, *Wonder Confronts Certainty: Russian Writers on the Timeless Questions and Why Their Answers Matter* (Cambridge, MA: Harvard University Press, 2023), 188–89.

Chapter 7

1 This was the legendary fifth Solvay Conference in Physics: see Image 7.

2 See Gilbert King, "Fritz Haber's Experiments in Life and Death," *Smithsonian Magazine*, June 6, 2012. Cf. Benjamín Labatut, *When We Cease to Understand the World*, trans. Adrian Nathan West (London: Pushkin Press, 2020), 27–29.

3 See Guido Bacciagaluppi and Antony Valentini, *Quantum Theory at the Crossroads: Reconsidering the 1927 Solvay Conference* (Cambridge, UK: Cambridge University Press, 2009), xv, 27 with n. a.

4 See Walter Moore, *Schrödinger: Life and Thought* (Cambridge, UK: Cambridge University Press, 1989), 197–201 (Kindle version).

5   Erwin Schrödinger, "The Present Situation in Quantum Mechanics" (1935), trans. John D. Trimmer in *Proceedings of the American Philosophical Society* 124 (1980): 323–38. See Image 8.

6   Galileo Galilei, *The Assayer*. See *The Essential Galileo*, ed. and trans. Maurice A. Finocchiaro (Indianapolis, IN: Hackett, 2008), 179–89.

7   Maxwell, "Molecules," *Nature* 8 (1873): 441. See further Abraham Pais, *Subtle is the Lord: The Science and the Life of Albert Einstein* (Oxford: Oxford University Press, 1982), 82.

8   Werner Heisenberg, *Physics and Beyond: Encounters and Conversations* (New York: Harper, 1971), 61.

9   Niels Bohr, "On the Constitution of Atoms and Molecules," *The Philosophical Magazine* 26 (1913): 1–25, 476–502, 857–75, esp. 12–14 and 476–502.

10   Niels Bohr, "Diskussion mit Einstein über erkenntnistheoretische Probleme in der Atomphysik," 1949, in *Albert Einstein als Philosoph und Naturforscher*, ed. P. A. Schilpp (Stuttgart: W. Kohlhammer Verlag, 1955), 115–50. Translated in *Niels Bohr: Collected Works*, vol. 6, ed. (Amsterdam: Elsevier, 2008), 349.

11   One of the clearest explanations of the uncertainty principle is to be found in Erwin Schrödinger, "Conceptual Models in Physics and their Philosophical Value," trans. W. H. Jonston in *Lectures on Physics and the Nature of Scientific Knowledge*, ed. Vesselin Petkov (Montreal, Canada: Minkowski Institute Press, 1961), 91–96.

12   See Stefan Rozentall, *Niels Bohr: Memoirs of a Working Relationship* (Copenhagen: Christian Ejlers, 1967), 103.

13   Those inherent limits in human knowledge and computation are the subject of detailed and rigorous examination by mathematician Stephen Wolfram on his blog: see "Computational Foundations for the Second Law of Thermodynamics," in *Writings*. 2023. https://writings.stephenwolfram .com/2023/02/computational-foundations-for-the-second-law-of-thermo dynamics/. In *The World Behind the World: Consciousness, Free Will, and the Limits of Science* (New York: Avid Reader Press, 2023), 200–205, Erik Hoel takes Wolfram's work (and the probabilistic indeterminacy in thermodynam- ics which eventually helped prompt the discovery of quantum mechanics) as a justification for saving free will from the deterministic universe of Laplace.

14   See Naum S. Kipnis, *The History of the Principle of the Interference of Light* (Basel: Springer, 1991), 65; Niels Bohr, "Diskussion mit Einstein über erken- ntnistheoretische Probleme in der Atomphysik," 1949, in *Albert Einstein als*

Philosoph und Naturforscher, ed. P. A. Schilpp (Stuttgart: W. Kohlhammer Verlag, 1955), 115–50. Translated in Niels Bohr: Collected Works, vol. 6, ed. Finn Aaserud (Amsterdam: Elsevier, 2008), 354–56.

15   See Leonard Susskind and Art Friedman, *Quantum Mechanics: The Theoretical Minimum: What You Need to Know to Start Doing Quantum Physics* (New York: Basic Books, 2014), 116–19, and cf. David Deutsch, *The Beginning of Infinity* (London: Penguin, 2011), 286–87, for a diagram, though Deutsch adopts an untenable explanation based on multiverses, on which see Philip Ball, *Beyond Weird: Why Everything You Thought You Knew about Quantum Physics Is Different* (London: Vintage, 2018), 92–93, 134–36, 109, 291–305.

16   Albert Einstein, *Relativity: The Special and General Theory, 100th Anniversary Edition*, with commentaries and background material by Hanoch Gutfreund and Jürgen Renn (Princeton: Princeton University Press, 2015), 178.

17   Erwin Schrödinger, "Indeterminism in Physics," trans. W. H. Jonston in *Lectures on Physics and the Nature of Scientific Knowledge*, ed. Vesselin Petkov (Montreal, Canada: Minkowski Institute Press, 1961), 35–36.

18   James Murphy, "Biographical Introduction," in *Lectures on Physics and the Nature of Scientific Knowledge*, edited by Vesselin Petkov (Montreal, Canada: Minkowski Institute Press, 1961), xvii–xviii.

19   Further experiment continues to indicate that some sort of mind, or consciousness, is the *only* thing that can definitively resolve quantum indeterminacies—there is no known purely mechanical process that can do so. See George Musser, *Putting Ourselves Back in the Equation* (New York, Farrar, Straus and Giroux, 2023), 12, 106–7.

20   See further Brandon Rickabaugh and J. P. Moreland, *The Substance of Consciousness: A Comprehensive Defense of Contemporary Substance Dualism* (New Jersey: Wiley Blackwell, 2024), 318–22.

Chapter 8

1   Francis Bacon, *The Historie of Life and Death* (1638), "To the Reader." See *The Works of Francis Bacon: Translations of the Philosophical Works 2*, ed. James Spedding, Robert Leslie Ellis, and Douglas Denon Heath (Cambridge, UK: Cambridge University Press, 2013), vol. 5. See further Glenn Ellmers, "Pandemic Pandemonium," *Claremont Review of Books* Summer 2023. https://claremontreviewofbooks.com/pandemic-pandemonium/.

2   Edge, "Death Is Optional: A Conversation: Yuval Noah Harari, Daniel Kahneman," March 4, 2013. Archived June 3, 2022. https://archive.is/ZGmrQ.

https://www.edge.org/conversation/yuval_noah_harari-daniel_kahneman -death-is-optional.

3   Augustine, *Confessions*, 10.6(9).

4   Plato, *Timaeus* 45b-d. See also Empedocles in *Die Fragmente der Vorsokratiker*, ed. Hermann Alexander Diels and rev. Walther Kranz (Berlin: Weidmannsche Buchhandlung, 6th ed. 1952), DK 31 B84 and 68, with Aristotle, *De Sensu et Sensato* 437b, Theophrastus, *De Sensibus* 7–8, and discussion in Jackson P. Hershbell, "Empedoclean Influences on the Timaeus," *Phoenix* 28, no. 2 (1974): 145–66, 157–58.

5   See Abdelghani Tbakhi and Samir S. Amr, "Ibn Al-Haytham: Father of Modern Optics," *Annals of Saudi Medicine* 27.6 (2007), 464. https://www .ncbi.nlm.nih.gov/pmc/articles/PMC6074172/; Abdelhamid I. Sabra, "Ibn Al-Haytham, Abu Ali Al-Hasan Ibn Al-Hasan," in *Dictionary of Scientific Biography*, vol. 6, ed. Gillespie Charles Coulston (New York: Charles Scribner's Sons, 1974), 189–210.

6   See Ibn al-Haytham, *Optics* I.4.1–28, I.6.56–61 in *The Optics of Ibn Al-Haytham Books I–III: On Direct Vision*, translated by Abdlhamid I. Sabra (London: The Warburg Institute, 1989), 51–55, 80–82. Cf. Galen, *De Usu Partium* X.12 in *Oevres anatomiques, physiologiques et médicales de Galien*, trans. Charles Daremberg, vol. 1 (Paris: J.-B. Ballière, 1854), 639.

7   See Aristotle, *De Anima* 425b and cf. Mark Eli Kalderon, *Form without Matter: Empedocles and Aristotle on Color Perception* (Oxford: Oxford University Press, 2015), 6–16, and Roland Polansky, *Aristotle's De Anima: A Critical Commentary* (Cambridge, UK: Cambridge University Press, 2007), 264.

8   See Ibn al-Haytham, *Optics* II.3.1–235 in *The Optics of Ibn Al-Haytham Books I–III: On Direct Vision*, trans. Abdelhamid I. Sabra (London: The Warburg Institute, 1989), 127–207 with David C. Lindberg, "Alhazen's Theory of Vision and its Reception in the West," *Isis* 58, no. 3 (1967): 323, 335; Nader El-Bizri, "A Philosophical Perspective on Alhazen's Optics," *Arabic Sciences and Philosophy* 15 (2005), 191–2.

9   For modern comments in the same vein, see Konrad Lorenz, *Studies in Animal and Human Behavior*, vol. 2, trans. Robert Martin (Cambridge, MA: Harvard University Press, 1970–1971), xxi–xxiii; Iain McGilchrist, *The Matter with Things: Our Brains, Our Delusions, and the Unmaking of the World* vol. 1 (London: Perspectiva Press, 2021), 253–54.

10 Parmenides in the so-called "On Nature" in Simplicius, *Physics* 117.2 (fragment 6), lines 11–16.

11   Plato, *Republic* 7.529b-c.

12   See pp. 52–56 and 142–43 on Galileo Galilei, *The Assayer*. See further *The Essential Galileo*, ed. and trans. Maurice A. Finocchiaro (Indianapolis, IN: Hackett, 2008), 179–89; Descartes, *Principia Philosophiae* II.4; *The World* chapter 6; and Francis Bacon, *Sylva Sylvarum* in *The Works of Francis Bacon*, ed. Robert Ellis and James Spedding (London: H. Bryer, 1803), I.98.

13   Newton, *Opticks* (New York, Dover Editions, based on the 4th edition, 1730), 26–54.

14   John Keats, "Lamia" II. 234–37. See further M. H. Abrams, *The Mirror and the Lamp: Romantic Theory and the Critical Tradition* (Oxford: Oxford University Press, 1953), 303–12.

15   Owen Barfield, *Saving the Appearances: A Study in Idolatry* (Middletown, Connecticut: Wesleyan University Press, 1965), 57.

16   Ibid., 55–56.

17   Iain McGilchrist, *The Matter with Things: Our Brains, Our Delusions, and the Unmaking of the World*, vol. 2 (London: Perspectiva Press, 2021), 1063. Cf. Carlo Rovelli, *Helgoland: Making Sense of the Quantum Revolution*, trans. Erica Segre and Simon Carnell (New York: Riverhead Books, 2020), 128, reflecting on the insights of Ernst Mach and Alexander Bogdanov.

18   Johann Wilhelm von Goethe, *Faust* II v.7.12104–7.

19   Johann Peter Eckermann, *Conversations with Goethe* (Gespräche mit Goethe), trans. John Oxenford (London: Dent, 1970), 415; entry for 20 June, 1831.

20   Johann Wilhelm von Goethe, *Theory of Color* (1810), trans. Charles Lock Eastlake (London: John Murray, 1840), xvii–xviii. See further xlvi, 286–89, §§ 722–29.

21   George Berkeley, *Three Dialogues between Hylas and Philonous* (1713), ed. Robert Merrihew Adams (Indianapolis, IN: Hackett, 1979), 25. See further *Of the Principles of Humane Knowledge* (1710) (London: Jacob Tonson, 1734), I.iv.

22   Martine Rothblatt, *From Transgender to Transhuman: A Manifesto on the Freedom of Form* (self-published, 2011), locations 845–50, Kindle edition.

23   Raymond Kurzweil, "The Law of Accelerating Returns," in the Kurzweil library + collections (March 7, 2001), https://www.thekurzweillibrary.com/the-law-of-accelerating-returns.

24   Edge, "Death is Optional: A Conversation: Yuval Noah Harari, Daniel Kahneman," March 4, 2013. Archived June 3, 2022. https://archive.is/ZGmrQ .https://www.edge.org/conversation/yuval_noah_harari-daniel_kahneman-death-is-optional.

25 See further Philip Ball, *Beyond Weird: Why Everything You Thought You Knew about Quantum Physics Is Different* (London: Vintage, 2018), 210–12.

26 See Hans Lewy, *Chaldean Oracles and Theurgy* (Paris: Institut d'Études Augustiniennes, 1978, revised edition), 151–52; Peter Kingsley, "Empedocles' Sun," *The Classical Quarterly* 44, no. 2 (1994), 316–24, esp. 316 and 323. The insight was not limited to the West; summarizing the *Vaiśeṣika Sūtra* of the ancient (c. sixth to second century BC) Indian philosopher Kaṇāda, Subhash Kak writes that "light has outer and inner aspects. The outer light is generated by atoms ... while the inner light is a consequence of consciousness." Subhash Kak, *The* Vaiśeṣika Sūtra *of Kaṇāda* (Ontario: Mount Meru Publishing, 2016), *Vaiśeṣika Sūtra* 44, on VS 8.1.2.

27 Jakob Böhme, *The Signature of All Things*, trans. John Ellistone in *The Works of Jacob Behmen, the Teutonic Philosopher* vol. 4 (London: M. Richardson, 1794), IV.43.

28 Thomas Traherne, *Centuries* (1600s), with an introduction by Michael Martin (Brooklyn, NY: Angelico Press, 2020), 88–89.

Chapter 9

1 Immanuel Kant, *Allgemeine Naturgeschichte und Theorie des Himmels* (Leipzig: Johann Petersen, 1755), 23–37; Pierre-Simon Laplace, *Exposition du système du Monde* (1796), 6th ed. (Paris: Gautier-Villars, 1835), 498–509.

2 See Duncan Aikman, "Lemaitre Follows Two Paths to Truth," *New York Times Magazine* (February 19, 1933), p. 3.

3 Lemaître's discovery of this fact predated that of Edwin Hubble, who became famous for it and gave his name to the constant which measures the universe's expansion. There has been some controversy over whether Lemaître's discovery was redacted in the English translation, giving Hubble apparent primacy. See Sidney van den Bergh, "The Curious Case of Lemaître's Equation No. 24," *Journal of the Royal Astronomical Society of Canada* 105, no. 4 (August 2011): 151. https://articles.adsabs.harvard.edu/pdf/2011JRASC.105.151V.

4 Georges Lemaître, "L'Hypothèse de l'atome primitive," *Revue des Questions scientifiques* 119 (1948): 339.

5 The astronomer Fred Hoyle, who argued against theories like Lemaître's in favor of a "steady-state" cosmology where the universe existed eternally, came up with the name "big bang" to describe what he opposed. But he did not do so in the derisive spirit usually attributed to him. See Helge Kragh, "How did the Big Bang Get its Name? Here's the Real Story," *Nature* 25 March 2024, https://www.nature.com/articles/d41586-024-00894-z.

6   Georges Lemaître, "Rencontres avec Einstein," *Revue des Questions scientifiques* 129 (1958): 129. See also Adam Curtis, "A Mile or Two off Yarmouth," *BBC Blogs*, February 24, 2012. https://www.bbc.co.uk/blogs/adamcurtis /entries/512cde83-3afb-3048-9ece-dba774b10f89.

7   Georges Lemaître, "Univers et Atome" (1963) in *L'itinéraire spirituel de Georges Lemaître* (Brussels: Lessius, 2007), 210–12. See further Thomas Hertog, *On the Origin of Time* (New York: Bantam Books, 2023), 35–36.

8   Stephen Hawking and Leonard Mlodinow, "Why God Did Not Create the Universe," *Wall Street Journal* September 30, 2010. https://www.wsj.com /articles/SB10001424052748704206804575467921609024244.

9   See Roger Penrose, *The Emperor's New Mind* (Oxford: Oxford University Press, 1989), 341–44; Stephen C. Meyer, *Return of the God Hypothesis* (New York: HarperOne, 2021), 200–252.

10 Werner Heisenberg, *Physics and Philosophy* (New York: Harper & Brothers, 1958), 41.

11 See p. 170 and George Musser, *Putting Ourselves Back in the Equation* (New York, Farrar, Straus and Giroux, 2023), 111–12. On the inescapability of conscious observation as an ingredient even in normal quantum phenomena, however, see Fritz London and Edmond Bauer, "La théorie de l'observation en mécanique quantique" (1939), trans. as "The Theory of Observation in Quantum Mechanics" in *Quantum Theory and Measurement*, ed. John Archibald Wheeler and Wojciech Hubert Zurek (Princeton: Princeton University Press, 1982), 250–256. See further Heinrich Päs, *The One: How an Ancient Idea Holds the Future of Physics* (New York: Basic Books, 2023), 250–66.

12 The theory summarized in Don Lincoln, "'Nothing' doesn't exist. Instead there is 'quantum foam,'" *The Big Think*, February 16, 2023 (https://archive.ph /SHaLp), that nothingness is actually occupied by "quantum foam"—i.e. the flicker of nonzero energy and matter out of zero kelvin states—is unpersuasive for the same reason that a purely mechanical big bang is unpersuasive. For the energy of the foam to be more than mere potential, and for the laws that govern it to select one out of its many possible timelines, something other than the foam itself—something other than material reality—must have made contact with it.

13 Bryce S. DeWitt, "Quantum Mechanics and Reality," *Physics Today* 23 (1970): 30–35, 34.

14 See Hugh Everett III, *Theory of the Universal Wave Function* (Thesis, Princeton University, 1956), 115–19.

246 LIGHT OF THE MIND, LIGHT OF THE WORLD

15 Schrödinger himself hinted at this answer: see *Interpretation of Quantum Mechanics: Dublin Seminars (1949–1955) and Other Unpublished Essays*, ed. Michel Bitbol (Oxford: Oxbow, 1955), 19–37.

16 See Peter Byrne, *The Many Worlds of Hugh Everett III: Multiple Universes, Mutually Assured Destruction, and the Meltdown of a Nuclear Family* (Oxford: Oxford University Press, 2010), 109–14, 137–38, 216–21; Stefano Osnaghi, Fabio Freitas, and Olival Freire Jr. "The Origin of the Everettian Heresy," *Studies in History and Philosophy of Modern Physics* 40, no. 2 (2009): 97–123.

17 See Laura Mersini-Houghton, *Before the Big Bang: The Origin of Our Universe from the Multiverse* (New York: Mariner, 2022), 128–69. Mersini-Houghton believes that "scarring" or irregularities in the cosmic microwave background furnish evidence of initial contact between our universe and others. Besides the contradiction in terms this implies, in fact the evidence neither does nor can support this conclusion with any certainty. See William H. Kinney, "Limits on Entanglement Effects in the String Landscape from Planck and BICEP/Keck Data," *Journal of Cosmology and Astroparticle Physics* 11 (2016): 1–21. https://arxiv.org/pdf/1606.00672.pdf.

18 See David Deutsch, *The Beginning of Infinity: Explanations that Transform the World* (London: Penguin, 2011), 258–303.

19 Compare the similar argument made in Stephen Hawking, *A Brief History of Time* (New York: Bantam, 1988), 176–81.

20 Philip Ball, "The Many Worlds Fantasy," *iai News* 96 (2021): https://iai.tv /articles/the-many-worlds-fantasy-auid-1793. See further Philip Ball, "Why · the Many-Worlds Interpretation Has Many Problems," *Quanta Magazine*, October 18, 2018 https://www.quantamagazine.org/why-the-many-worlds -interpretation-has-many-problems-20181018/; George Ellis and Joe Silk, "Scientific Method: Defend the Integrity of Physics," *Nature* 516 (2014): 321–23. https://www.nature.com/articles/516321a; George Ellis, "Does the Multiverse Really Exist?" *Scientific American* (2011): https://www.scientificamerican.com /article/does-the-multiverse-really-exist/; Stephen C. Meyer, *Return of the God Hypothesis* (New York: HarperOne, 2021), 507; Spencer A. Klavan, "Worlds Without End," *The Claremont Review of Books* (Summer 2022): https://claremontreviewofbooks.com/worlds-without-end/.

21 Cf. Thomas Hertog, *On the Origin of Time* (New York: Bantam Books, 2023), 133–64.

22 See Steven Weinberg, *Cosmology* (Oxford: Oxford University Press, 2008), 101–32, with Image 9.

23 See NASA, "Background on the Background Explorer and the Science of John Mather," Nasa.gov, October 5, 2006. https://archive.ph/8Drx.

24 See further Harry Baker, "Rare Interview with 'Father of the Big Bang' Rediscovered after 60 Years," Space.com, February 6, 2023. https://www.space .com/lost-georges-lemaitre-big-bang-interview-recovered https://archive.ph /MhYtP, and Image 10.

25 E.g. Thomas Aquinas, *Summa Theologica* Part I question 74, and perhaps Basil, *Hexaemeron* 2.8, though the latter seems to leave open the possibility that the author of Genesis wrote "one day" instead of "the first day" to mark the period off as a unique "eternal instant." R. C. Sproul makes the modern case for a young earth, but he also gives a helpful and fair summary of various other modern interpretations in R. C. Sproul, *Truths We Confess: A Systematic Exposition of the Westminster Confession of Faith* (Orlando, FL: Reformation Trust, 2019), 99–114. See further Rodney Stark, *Bearing False Witness: Debunking Centuries of Anti-Catholic History* (West Conshohocken, PA: Templeton Press, 2016), 89.

26 See for example Irenaeus, *Against Heresies* 8.28.3, Philo of Alexandria, *On the Creation of the World* 3. Cf. Michael Roberts, "Genesis Chapter 1 and Geological Time from Hugo Grotius and Marin Mersenne to William Conybeare and Thomas Chalmers (1620–1825)," presented at the Science and Belief conference, Durham, 2002. https://michaelroberts4004.files.wordpress .com/2018/09/genesis-1-geological-time-from-1600–1850.pdf; Howard Van till, "Basil Augustine, and the Doctrine of Creation's Functional Integrity," *Science and Christian Belief* 8 (1996): 21–38. For subtle modern reflections on the slippage between divine and human time see Jorge Luis Borges, "A History of Eternity" (1936) in *Selected Non-Fictions*, ed. Eliot Weinberger (New York: Penguin, 2000), 2000.

27 Augustine, *City of God* 11.7. See translation by Marcus Dods in *Nicene and Post-Nicene Fathers*, first series, vol. 2, ed. Philip Schaff (Buffalo, NY: Christian Literature Publishing Co., 1887). Revised and edited for the website New Advent by Kevin Knight. http://www.newadvent.org/fathers/120111.htm. Cf. *De Genesi ad Litteram Imperfectus Liber* 6.27–7.28 at https://www.augustinus .it/latino/genesi_incompiuto/index.htm and *De Genesi ad Litteram Libri Duodecim* 1.12. 24–25 at https://www.augustinus.it/latino/genesi_lettera /index2.htm.

28 John Wheeler, "Genesis and Observership" in *Foundational Problems in the Special Sciences: Part Two of the Proceedings of the Fifth International Congress*

of Logic, Methodology, and Philosophy of Science, ed. R. E. Butts and J. Hintikka (Dordrecht: D. Reidel Publishing Company, 1975), 3–34, 6, 24–25.

29   Thomas Hertog, *On the Origin of Time* (New York: Bantam Books, 2023), 196.

30   See Adam Frank and Marcelo Gleiser, "The Story of Our Universe May be Starting to Unravel," *New York Times*, September 2, 2023. https://archive.ph/ gHBVp; Ben Turner, "James Webb Telescope Confirms there is Something Seriously Wrong with our Understanding of the Universe," *LiveScience* March 14, 2024. https://www.livescience.com/space/cosmology/james-webb-telescope -confirms-there-is-something-seriously-wrong-with-our-understanding-of- the-universe. For the astounding imagery associated with these discoveries see Jay Bennett, "Time Travel to Ancient Galaxies" in *National Geographic: The Space Issue* (October 2023), 86–112; and *The SkyLive* space map: https: //theskylive.com/planetarium?obj=jwst. See Images 11–13.

31   See Rajendra P. Gupta, "JWST early Universe observations and ΛCDM cos- mology," *Monthly Notices of the Royal Astronomical Society* 524, no. 3: September 2023, 3385–95. https://arxiv.org/abs/2309.13100. But ct. Evrim Yazgin, "Is the Universe Twice as Old as We Thought?" *Cosmos*, July 19th, 2023. https://www.astronomy.com/science/is-the-universe-twice-as-old-as-we -thought/.

32   Augustine, *Confessions* 11.27(36). See translation by Carolyn J.-B. Hammond in *Augustine: Confessions Books 9–13* (Harvard: Loeb Classical Library, 2016), 249.

Chapter 10

1   See for instance Michael Shermer's summary of "Quantum Flapdoodle" in Victor J. Stenger, *Quantum Gods: Creaton Chaos, and the Search for Cosmic Consciousness* (New York: Prometheus Books, 2009), 7–9. See further Stenger's 14–15, 55–61.

2   Michel Foucault, interview, "Truth and Power" in *Power/Knowledge: Selected Interviews and Other Writings 1972–1977*, ed. Colin Gordon (New York: Pantheon, 1980), 131.

3   This conviction—known as "scientism"—is the natural offspring of objectiv- ism and the governing superstition of our era. See Aaron Kheriaty, "Technocracy and Totalitarianism" *The American Mind*, November 17, 2022. https://americanmind.org/features/technocracy-and-totalitarianism/.

4   See further Owen Barfield, *Saving the Appearances: A Study in Idolatry* (Middletown, Connecticut: Wesleyan University Press, 1965), 15–18.

5   For a fictional but highly suggestive proposition that subatomic particles may
    reach a degree of complexity in the eleven dimensions of string theory that
    is capable of encoding "wisdom" or "intelligence," see Cixin Liu, *The Three-
    Body Problem* (2004), trans. Ken Liu (New York: Tor, 2019), 366.
6   Words, or "signs," not only name objects but gather and organize sense per-
    ceptions into entities that we can experience as having a distinctive character
    or form: this is a rich area of study in the field of semiotics. See esp. Walker
    Percy, *Lost in the Cosmos: The Last Self-Help Book* (New York: Open Road
    Media, 1983), Kindle edition, locations 1468–75; Philip E. L. Smith, *Cro-
    Magnon Man*, ed. Tom Prideaux (New York: Time-Life Books, 1973), 7;
    Ferdinand de Saussure, *Course in General Linguistics*, ed. Charles Bally and
    Albert Sechehaye, trans. Wade Baskin (New York: Philosophical Library, 1950),
    107–26; Jerrold J. Abrams, "Peirce, Kant, and Apel on Transcendental
    Semiotics: The Unity of Apperception and the Deduction of the Categories
    of Signs," *Transactions of the Charles S. Peirce Society* 40 (2004), 627–77.
7   Aquinas, *De Natura Verbi Intellectus* Ch.1. Cf. Richard M. Weaver, *Ideas Have
    Consequences* (1948) (Chicago: University of Chicago Press, expanded edition,
    2013), 134–35.
8   G. K. Chesterton, *The Everlasting Man* (Dehli: Grapevine, 1925), Kindle edi-
    tion, p. 16. Cf. Michael Polanyi, *Personal Knowledge: Towards a Post-Critical
    Philosophy* (Chicago: The University of Chicago Press, 1958), 387–90.
9   See Walker Percy, *Lost in the Cosmos: The Last Self-Help Book* (New York:
    Open Road Media, 1983), Kindle edition, locations 1095–1174. On the differ-
    ence between human language and exchanges of vocal signals between ani-
    mals, see Ludwig Wittgenstein, *Philosophical Investigations* (1953), trans.
    G. E. M. Anscombe, P. M. S. Hacker, and Joachim Schulte (London: Wiley-
    Blackwell, 2009), 145 §493. On the first appearance of humanity through
    symbol and language, see Hans Jonas, *The Phenomenology of Life: Toward a
    Philosophical Biology* (Evanston, IL: Northwestern University Press, 1966),
    157–73; Brendan Purcell, *From Big Bang to Big Mystery: Human Origins in the
    Light of Creation and Evolution* (New York: New City Press, 1994), 173–85,
    208–39, esp. the discussion of cave paintings.
10 Rudolf Steiner, *Die Philosophie der Freiheit* (Luxembourg: Amazon Media
    EU, 1918), 81.
11 Diodorus Siculus, *Library of History*, fragments of Book 26.18. See *The Library
    of History of Diodorus Siculus*, vol. 11, trans. Francis R. Walton (Harvard: Loeb
    Classical Library, 1957), 176–201.

12   See Hayden Dunham, "Pippa Gardner and Hayden Dunham on the Struggle of Being Inside Bodies," *Interview Magazine*, May 19, 2021. https://archive.ph /e6skx.

13   "Political/Religious Arguments" in *The Antinatalism Argument Guide* (https: //antinatalismguide.wixsite.com/guide/political-religious-arguments).

14   The subreddit r/antinatalism has 213,000 members as of January 2024: https: //www.reddit.com/r/antinatalism/. See further Rebecca Tuhus-Dubrow, "I wish I'd never been born: The Rise of the Antinatalists," in *The Guardian*, November 14 2019. https://archive.ph/Yy7FW; David Benatar, *Better Never to Have Been: The Harm of Coming into Existence* (Oxford: Oxford University Press, 2006). Hence Elon Musk frequently argues that "the real battle is between the extinctionists and the humanists." @elonmusk, October 31st, 2023. https://x.com/elonmusk/status/1719398881870356695?s=20. See further Cara Buckley, "Earth Now Has 8 Billion Humans. This Man Wishes There Were None," *New York Times*, November 23, 2023. https://www.nytimes .com/2022/11/23/climate/voluntary-human-extinction.html.

15   Martine Rothblatt, *From Transgender to Transhuman: A Manifesto on the Freedom of Form* (self-published, 2011), kindle edition, location 850. See further Anonymous, "Gender Acceleration: A Blackpaper" in The Anarchist Library, October 31, 2018. https://archive.ph/MkJk8.

16   Samuel Taylor Coleridge, *Biographia Literaria* (1817), ed. Adam Roberts (Edinburgh: Edinburgh University Press, 2014), 217–18. See further M. H. Abrams, *The Mirror and the Lamp: Romantic Theory and the Critical Tradition* (Oxford: Oxford University Press, 1953), 292–93.

17   See Egert Pöhlmann and Martin L. West, *Documents of Ancient Greek Music* (Oxford: Oxford University Press, 2001), 191–93.

18   "This is the Feast of Victory for Our God," composed by Richard Hillert, lyrics translated from *Dignus est Agnus* (Revelation 5:12–13, 19:5–9) by John W. Arthur (1970). Archived at hymnary.org: https://archive.is/en6US (originally featured in the *Lutheran Book of Worship*). Cf. e.g. David Blackwell, *Scottish Blessing* (St. Louis: Morningstar Music Publishers, 2014).

19   John Archibald Wheeler, "Information, Physics, Quantum: The Search for Links," in *Proceedings of the 3rd International Symposium in the Foundations of Quantum Mechanics*, ed. Hiroshi Ezawa, Shun Ichi Kobayashi, and Yoshimasa Murayama (Tokyo: Physics Society of Japan, 1989), 354–68.

20   Eleanor Farjeon, "Morning Has Broken" (1931), archived at hymnary.org: https://archive.is/FZ205 (first collected in *Songs of Praise*, 2nd ed). See further Michael Martin, *Sophia in Exile* (Angelico Press: New York, 2021), 95–110.

21 Thomas Traherne, *Centuries* (1600s), with an introduction by Michael Martin (Brooklyn, NY: Angelico Press, 2020), 44–45.

22 C. S. Lewis, *The Four Loves* (New York: HarperCollins, 1960), 79. The old author to whom Lewis refers is the one known as Dionysius the Areopagite, in *Celestial Hierarchy* 10.273a-b. See further Brendan Purcell, *From Big Bang to Big Mystery: Human Origins in the Light of Creation and Evolution* (New York: New City Press, 1994), 297–300.

Afterword: An Invitation

1 Charles Haddon Spurgeon, "Knowledge Commended," from Metropolitan Tabernacle Pulpit, volume 11, January 15, 1865. Reprinted by The Spurgeon Center for Biblical Preaching at Midwestern Seminary https://www.spurgeon .org/resource-library/blog-entries/spurgeon-darwin-and-the-question-of -evolution/#_ftn6.

2 Albert Einstein, "Notes for an Autobiography," *Saturday Review of Literature* (November 26, 1949), 9–12. See further Walter Isaacson, *Einstein: His Life and Universe* (New York: Simon & Schuster, 2007), 18–21.

3 "Down with God! How the Soviet Union took on religion—in pictures," *The Guardian*, October 23, 2019. https://www.theguardian.com/artanddesign /gallery/2019/oct/23/down-with-god-how-the-soviet-union-took-on-religion -in-pictures.

4 Alfred, Lord Tennyson, "*In Memoriam* A. H. H. Obiit MDCCCXXXIII," 119.7–12.

5 George Berkeley, *Three Dialogues between Hylas and Philonous* (1713), ed. Robert Merrihew Adams (Indianapolis, IN: Hackett, 1979), 88.

Index